BOISE FOOTHILLS AND FOREST PLANTS

including
Boise Peak, Bogus Basin,
and Mores Mountain

BOISE FOOTHILLS AND FOREST PLANTS

including
Boise Peak, Bogus Basin,
and Mores Mountain

by Ray S. Vizgirdas

Illustrations by
Edna M. Rey-Vizgirdas

First Printing: 2017

ISBN # 978-1-387-73674-4

Mountainforaging@gmail.com

Disclaimer

While this book documents the uses of wild plants found in the Boise Foothills and Forested areas, the author disclaims any liability for injury that may result from following any instructions for collecting, preparing or consuming plants described in this guide. Efforts have been taken to assure the descriptions of plants represented are accurate representations of the family, genus, and species noted. It should be understood that growth conditions, improper identification, and varietal differences, as well as an individual's own sensitivity or allergic response can contribute to a hazard in sampling or using a plant. Furthermore, the reader is encouraged to seek the assistance from experienced botanists in identifying any of the plants discussed in this book.

Acknowledgments

There are several individuals that have been supportive in one way or another, as well as providing encouragement over the many years. I say to you THANKS - I truly have appreciated your guidance: Dr. Mark Plew, Dr. Roger Rosentretter, Ms. Kay Beall, Ms. Ann Debolt, Dr. Kerry McClay, Dr. Jim Smith, Dr. John Cossel, Dr. Lynn Kinter, Dr. Don Mansfield, Ms. Jennifer Miller, and Mr. Michael Mancuso. Of course, the most important person I must thank is my wife Edna - for being patient and at the same time cracking the whip.

INTRODUCTION

The aim of this handbook is to provide a guide to the common vascular plants of the Boise Foothills, including the forested area from Boise Peak westward to Bogus Basin Ski Area and Mores Mountain. This book provides information on the natural history, ecology, and adaptations of common trees, shrubs, and wildflowers, as well as how the plants were used by Native Americans and early inhabitants of the area.

Please keep in mind that this is not intended to be a "how-to" book and the descriptions of plant usage do not advocate nor encourage experimentation by the reader. Many plants have toxic properties and should not be used without thorough knowledge of that particular plant species. It is hoped that this book will be of interest to many different segments of the population that visit the Boise Foothills: the hikers and backpackers; biologists; botanists; ethnologists; archeologists; outdoor and environmental educators; naturalists; outdoor recreationists; park rangers; and members of the general public.

The area covered by this guide extends from approximately Lucky Peak Reservoir westward to Highway 55, and from the lower reaches from Camelsback Park and Fort Boise up to Boise Peak across to Bogus Basin and Mores Mountain. Habitats included are sagebrush scrub, grassland, willow riparian, ponderosa pine forest, and anthropogenic.

The plant life in the Boise Foothills can best be described as *depauperate*. From a biological perspective that means "impoverished" or "having limited biodiversity." Being so close to Boise, the foothills have experienced numerous impacts over many years. Impacts include historic and current livestock grazing, housing developments, more frequent wild fires, and recreation (hiking, biking, trail construction). These impacts have allowed numerous non-native species of plants to establish themselves in the foothills and effect adverse changes. *However, patches and stringers of habitat containing native plants and animals can still be found.*

A Word on Weeds

Some of the plants covered in this book include species that are not native - variously called "weeds, invasives, and exotics." A few of these are descended from garden plants introduced in the early years of settlement in the area or having recently escaped from housing developments making their way into the foothills. Many have seeds that are easily carried in on the shoes and tires, and by the wind. For the most part, these species are restricted in their distribution, growing in places where disturbance of the soil is routine, often following roadsides and trails. The species of greatest concern are those that have the ability to aggressively colonize and push aside native vegetation (e.g., cheatgrass [*Bromus tectorum*] and knapweed [*Centaurea*]).

Furthermore, a number of native species can also behave in a "weedy" manner. Often annuals, they have the ability to invade and maintain themselves on ground that is subjected to disturbances, especially at roadsides and along trails.

References The following references provide additional information about the natural history, ecology, and ethnobotany of Boise Foothill plants. These books contain more comprehensive lists of references and literature citations that were used to create this guide. A robust list of references is not provided here simply because we wanted to keep it a pocket-sized guide.

Vizgirdas, R.S. 2017. *Wild Edible and Useful Plants of Idaho.* RSVizgirdas through Lulu.com

Vizgirdas, R.S. 2017. *Field Guide to South-central Idaho Plants.* RSVizgirdas through Lulu.com

Vizgirdas, R.S. 2007. *A Guide to Plants of Yellowstone and Grand Teton National Parks.* University of Utah Press

Vizgirdas, R.S. 2003. *Useful Plants of Idaho.* Idaho State University Press.

KEY TO SOME OF THE COMMON PLANT FAMILIES IN THE BOISE FOOTHILLS

(please note, not all families are included here; a more technical flora will need to be consulted for some groups of plants)

1. Plants reproducing by spores; ferns and horsetails ----- **2**
1. Plants reproducing by seeds; typical flowering plants and gymnosperms ----- **3**

2. Stems hollow and jointed; leaves in whorls ----- **Horsetail Family (Equisetaceae)**
2. Stem solid; leaves fern-like ----- **Ferns**

3. Ovules and seeds borne on face of a scale, not enclosed in an ovary (e.g., fruits); evergreen tree ("pines") ----- **Gymnosperms in Pinaceae**
3. ovules and seeds in a fruit; typical flowering plants ----- **4**

4. Leaves usually parallel-veined; flower parts in threes or sixes ----- **5 ("Moncots")**
4. Leaves usually netted veined; flower parts in fours or fives; rarely in twos -----   **10 ("Dicots")**

5. Plants with petal-like perianth ----- **6**
5. Plants with perianth of chaffy scales or hairy bristles; rushes, grasses, sedges, and cattails ----- **8**

6. Ovary inferior ----- **7**
6. Ovary superior ----- **Lily Family (Liliaceae)**

7. Flower parts regular, alike in size and shape ----- **Iris Family (Iridaceae)**
7. Flower parts irregular, not alike ----- **Orchid Family (Orchidaceae)**

8. Plants growing in marshy places ----- **9**
8. Plants not confined to wet, marshy places ----- **Grass Family (Poaceae)**

9. Stems round or nearly so ----- **Cattail Family (Typhaceae)**
9. Stems appearing triangular because of the leaves in three rows ----- **Sedge Family (Cyperaceae)**

10. Petals absent ----- **11**
10. Petals present ----- **30**

11. Trees or shrubs ----- **12**
11. Herbaceous plants (maybe woody/shrubby at the base) ----- **16**

12. Flowers not in catkins ----- **13**
12. Flowers in catkins ----- **15**

13. Leaves opposite ----- **Maple Family (Aceraceae)**
13. Leaves alternate ----- **14**

14. Ovary with one cell ----- **Elm Family (Ulmaceae)**
14. Ovary with two or four cells ----- **Buckthorn Family (Rhamnaceae)**

15. Calyx present; ovary with one or two ovules or seeds; leaves without stipules ----- **Birch Family (Betulaceae)**
15. Calyx absent; ovary with many ovules or seeds; leaves usually with stipules ----- **Willow Family (Salicaceae)**

16. Leaves opposite ----- **17**
16. Leaves not opposite ----- **20**

17. Flowers perfect (having stamens and pistiles); plants without stinging hairs ----- **18**
17. Flowers not perfect; plant with stinging hairs ----- **Nettle Family (Urticaceae)**

18. Sepals united into a corolla-like tube ----- **Four-o'clock Family (Nyctaginaceae)**
18. Sepals not united, but free to base ----- **19**

19. Fruit a many-seeded capsule; leaves opposite -----**Pink Family (Caryophyllaceae)**
19. Fruit an achene; leaves seldom opposite ----- **Buckwheat Family (Polygonaceae)**

20. Flowers with stamens and pistils in separate flowers on the same plant; surface of plant usually scurfy or mealy ----- **Goosefoot Family (Chenopodiaceae)**
20. Flowers perfect, or if not perfect, then staminate and pistillate flowers are on separate plants; leaves not mealy ----- **21**

21. Pistils more than one ----- **22**
21. Pistils only one ----- **23**

22. Receptacle cup-shaped with stamens attached on the rim of the cup and above the ovary ----- **Rose Family (Rosaceae)**
22. Receptacle cone-like or flat, but never cup-shaped; stamens attached below or at base of the ovary ----- **Buttercup Family (Ranunculaceae)**

23. Ovary more than one-celled ----- **Mustard Family (Brassicaceae)**
23. Ovary only one-celled ----- **24**

24. Ovary wholly superior ----- **25**
24. Ovary partly inferior ----- **28**

25. Sepals and petals four each, stamens 6 ----- **Caper Family (Capparaceae)**
25. Sepals and petals not four each, stamens not six ----- **26**

26. Leaves with papery sheathing stipules, especially in young stems (except Eriogonum, no stipules) ----- **Buckwheat Family (Polygonaceae)**
26. Leaves without sheathing stipules ----- **27**

27. Fruit an achene; leaves not mealy ----- **Buckwheat Family (Polygonanceae)**
27. Fruit not an achene, but a utricle; leaves often mealy ----- **Goosefoot Family (Chenopodiaceae)**

28. Flowers on a leafy stem ----- **Sandalwood family (Santalaceae)**
28. Flowers on stem with few or no leaves ----- **Saxifrage Family (Saxifragaceae)**

30. Petals separate, not united to each other to form a tubular corolla ----- **31**
30. Petals united to each other at least at the base, to form a tubular corolla ----- **61**

31. Ovary superior ----- **32**
31. Ovary wholly or partly inferior ----- **56**

32. Stamens same number as the petals and opposite them ----- **33**
32. Stamens either not the same number as the petals, or alternate with them ----- **36**

33. Shrubs or small trees ----- **34**
33. Herbaceous plants ----- **35**

34. Leaves compound ----- **Barberry Family (Berberidaceae)**
34. Leaves simple ----- **Buckthorn Family (Rhamnaceae)**

35. Sepals 2; plants fleshy or succulent ----- **Purslane Family (Portulacaceae)**
35. Sepals 6; plants not succulent ----- **Barberry Family (Berberidaceae)**

36. Pistils one to many, but always simple, as indicated by only one stigma, style, or placenta ----- **37**
36. Pistil 1, but always compound, as shown by the two or more stigmas, styles, locules or placentae ----- **43**

37. Pistil one ----- **38**
37. Pistil 2 to many ----- **39**

38. Flowers irregular, papilionaceous ----- **Pea Family (Fabaceae)**
38. Flowers regular ----- **Rose Family (Rosaceae)**

39. Sepals not united to each other ----- **40**
39. Sepals united to each other to form a floral cup around the ovary ----- **41**

40. Stamens 10 ----- **Stonecrop Family (Crassulaceae)**
40. Stamens many (>10) ----- **Buttercup Family (Ranunculaceae)**

41. Stamens 5-12, usually 10 ----- **Saxifrage Family (Saxifragaceae)**
41. Stamens many ----- **42**

42. Stamens united into one or more series around the pistils ----- Mallow Family (Malvaceae)
42. Stamens free and distinct, not united ----- Rosaceae

43. Ovary deeply 5-lobed with one seed in each lobe; or ovaries united in a ring around a central axis ----- **44**
43. Ovary not appearing to be 5-lobed, nor united in a ring around the central axis ----- **45**

44. Stamen filaments united into a tube; carpels 5-9; ovaries united in a ring around a central axis ----- **Mallow Family (Malvaceae)**
44. Stamen filaments not united to each other ----- **Geranium Family (Geraniaceae)**

45. Sepals 2 ----- **Purslane Family (Portulacaceae)**
45. Sepals 4 to 5 ----- **46**

46. Sepals not united to each other ----- **47**
46. Sepals united to each other ----- **52**

47. Sepals 5; petals 5 ----- **48**
47. Sepals four; petals four ----- **Mustard Family (Brassicaceae)**

48. Flowers irregular ----- **Violet Family (Violaceae)**
48. Flowers regular ----- **49**

49. Stamens united with the base of the corolla; stamens more numerous than corolla lobes; ovary 3 to many celled ----- **Mallow Family (Malvaceae)**
49. Plants not as above ----- **50**

50. Flowers pink or red ----- **Geranium Family (Geraniaceae)**
50. Flowers white or blue ----- **51**

51. Flowers white ----- **Pink Family (Caryophyllaceae)**
51. Flowers blue -----**Flax Family (Linaceae)**

52. Leaves alternate or basal ----- **53**
52. Leaves opposite ----- **55**

53. Woody vines or shrubs ----- **Sumac Family (Anacardiaceae)**
53. Herbs ----- **54**

54. Stamens 10 or less ----- **Saxifrage Family (Saxifragaceae)**
54. Stamens numerous (>10) ----- **Rose Family (Rosaceae)**

55. Fruit a 2-winged samara ----- **Maple Family (Aceraceae)**
55. Fruit a capsule ----- **Pink Family (Caryophyllaceae)**

56. Flowers in umbels ----- **Carrot Family (Apiaceae)**
56. Flowers not in umbels ----- **57**

57. Stamens the same number as the petals and opposite them ----- **Saxifrage Family (Saxifragaceae)**
57. Stamens not the same number as the petals and alternate with them ----- **58**

58. Calyx lobes and petals five each ----- **59**
58. Calyx lobes and petals four each ----- **60**

59. Stamens many ----- **Rose Family (Rosaceae)**
59. Stamens 5-10 ----- **Saxifrage Family (Saxifragaceae)**

60. Shrubs ----- **Dogwood Family (Cornaceae)**
60. Herbs ----- **Evening-primrose Family (Onagraceae)**

61. Stamens more than 5 ----- **62**
61. Stamens five or less ----- **63**

62. Stamens many, their filaments united into a tube ----- **Mallow Family (Malvaceae)**
62. Stamens 6, united into two sets of 3 each ----- **Fumatory Family (Fumariaceae)** Dicentra needs to be included

63. Ovary superior ----- **64**
63. Ovary inferior ----- **76**

64. Corolla irregular (*Veronica* in Scrophulariaceae is slightly irregular) ----- **65**
64. Corolla not at all irregular ----- **68**

65. Leaves alternate or basal, or only the lower ones opposite ----- **66**
65. Leaves all opposite or whorled ----- **67**

66. Root parasites without chlorophyll or normal foliage ----- **Broom-rape Family (Orobanchaceae)**
66. Green plants with normal foliage, not parasites ----- **Figwort Family (Scrophulariaceae)**

67. Mature ovary deeply lobed or divided into four 1-seeded nutlets; style 2-branched; stems square; one seed per cell of the ovary ----- **Mint Family (Lamiaceae)**
67. Mature ovary a capsule, not lobed into nutlets, but may be notched so that it appears somewhat lobed; stems round; several seeds per cell of ovary ----- **Figwort Family (Scrophulariaceae)**

68. Ovary divided or lobed into 4 parts around the base of the style ----- **Borage Family (Boraginaceae)**
68. Ovary not 4-lobed around the style ----- **69**

69. Stamens opposite the corolla lobes ----- **Primrose Family (Primulaceae)**
69. Stamens attached between the corolla lobes (alternate) ----- **70**

70. Twinning or trailing herbs ----- **71**
70. Plants not twinning or trailing ----- **72**

71. Sepals separate to the base ----- **Morning Glory Family (Convolvulaceae)**
71. Sepals united ----- **Nightshade Family (Solanaceae)**

72. Ovary 1-celled ----- **73**
72. Ovary 2 or more celled ----- **74**

73. Leaves entire, opposite or whorled ----- **Gentian Family (Gentianaceae)**
73. Leaves, if entire, alternate or basal ----- **Waterleaf Family (Hydrophyllaceae)**

74. Stamens extending beyond the corolla ----- **Waterleaf Family (Hydrophyllaceae)**
74. Stamens shorter than the corolla ----- **75**

75. Style unbranched ----- **Nightshade Family (Solanaceae)**
75. Style branched ----- **Phlox Fmily (Polemoniaceae)**

76. Flowers in heads surrounded by several leaf-like bracts; anthers united in a ring or tube around the pistil ----- **Sunflower Family (Asteraceae)**
76. Flowers not as above ----- **77**

77. Shrubs or woody vines ----- **Honeysuckle Family (Caprifoliaceae)**
77. Herbs ----- **78**

78. Stamens 1-3; upper leaves apposite ----- **Valerian Family (Valerianaceae)**
78. Stamens 4 or 5; leaves whorled ----- **Madder Family (Rubiaceae)**

COMMON TREES, SHRUBS, AND WILDFLOWERS

MAPLE

Acer   ACERACEAE

General Description Maples are deciduous trees or shrubs with male and female flowers on the same or separate plants. Flowers are arranged in racemes, corymbs, or umbels. Fruits are winged schizocarps that resemble tiny "helicopters" when blown by the wind. Within the Boise Foothills two species may be encountered: Rocky Mountain maple (*A. glabrum*) and boxelder (*A. negundo*). Current research places the maple family in the mainly tropical Soapberry family (Sapindaceae).

Comparison of *Acer* species

A. glabrum Leaves simple
A. negundo Leaves compound with 3-5 leaflets

Ecology & Ethnobotany Like with other maples, syrup can be extracted by boiling down the sap. The winged seeds can be roasted for food. The young shoots and inner bark are valuable in times of emergency as food – dried and ground into flour. Maple wood has been used to make snowshoe framing, mush paddles, and other household utensils. Knots and burls on tree

trunks can be used for making bowls, dishes, and other items. Inner bark can be shredded and twisted into a coarse rope.

AMARANTH, PIGWEED
Amaranthus   AMARANTHACEAE

General Description In general, these are herbaceous annuals with small greenish flowers, and alternate entire or wavy margined leaves. There are at least 4 species of *Amaranthus* in the Boise Foothills.

Ecology & Ethnobotany The seeds of all species can be eaten whole as a cereal or ground into meal and made into cakes. The seeds are best collected in summer when the plants are fully mature. To free the seeds from their husks, rub the seed clusters between your hands. You can then winnow the seeds if there is a breeze, or if air is calm, slowly pour the seeds out of your hands and blow the chaff away. The seeds contain approximately 15 grams of protein per 100 grams, more than is found in rice and corn, and equal to, if not surpassing that found in wheat. When ground into a flour, amaranth has a distinctive flavor that is a bit strong used alone. We find it is better when mixed with other flours for breads and pancakes.

The highly nutritious amaranth contains more fiber and calcium than any other cereal grain in addition to a wide spectrum of vitamins and minerals, including Vitamins A and C, calcium, magnesium, and iron.

Amaranths is rich in the amino acid lysine, a product scarce in true cereal grains, thereby providing a more balanced source of protein.

The edible young shoots and leaves have a pleasant taste if eaten as a potherb soon after collection. Since the plants can accumulate nitrates, it is wise not to consume large quantities where nitrate fertilizers are used.

SKUNKBRUSH
Rhus aromatica　ANACARDIACEAE

General Description Skunkbush is a shrub that occurs on dry sunny slopes, has compound leaves with three leaflets, of which the middle leaflet is largest. Skunkbush has small yellow-green flowers that bloom before the leaves come out. The red-orange fruits are sour tasting. *Rhus aromatica* was previously classified as *Rhus trilobabta*.

A second sumac (*R. glabra*) occurs in the area. This native shrub is usually encountered in the lower elevations of the Boise Foothills.

Comparison of *Rhus* species

Skunkbrush (R. aromatica) Leaflets 3 or 5
Sumac (R. glabra) Leaflets more than 5

Ecology & Ethnobotany Skunkbush is one of the plants that shows measles-like splotching in the presence of ozone. As such, it is being watched by scientists in some areas as an indicator of air pollution.

The berries can be steeped in water to make an interesting tasting drink. The slender, flexible branches

of skunkbrush can be used for weaving baskets as they are somewhat vine-like. The branches can also be used as chew-sticks to clean teeth and massage gums. Take a small stem several inches long, remove the outer bark and chew on the tip to soften fibers. *Since some people may have an allergic reaction to the oils of sumac, it is recommended that this be done sparingly.*

POISON IVY
Toxicodendron rydbergii   ANACARDIACEAE

General Description Poison ivy is a low shrub or woody vine found in waste places, hillsides and rocky ravines in the Boise Foothills. The leaves are compound with three green, that are oval in shape, and pointed leaflets, which turn bright red or orange in the fall. The white flowers arise from the leaf axils, and the fruits are

white berries.

Ecology & Ethnobotany The foliage is poisonous, causing contact dermatitis. The actual skin irritant is found in the sap. The itchy or painful rash that develops from contact with the sap is greatest in spring and summer when the sap is abundant and the plant is easily bruised. Shortly after contact, the symptoms include itching, burning, and redness. Small blisters may appear after a few to several hours. Severe dermatitis, with large blisters and local swelling, can remain for several days and may require hospitalization. Secondary infections may occur when the blisters are broken. To help alleviate itching immediately, thorough washing with soap and water after contact is recommended.

BISHOP'S GOUTWEED
Aegopodium podagraria   APIACEAE

General Description This herbaceous, non-native perennial grows up to 3 feet tall. The basal and lower leaves have long petioles, and there are generally 9 leaflets per lower leaf. Each of the leaflets is ovate in shape and has an acute or acuminate apex. The bases of these leaflets can be rounded or cordate. The horticultural variety usually grown has white margins on its leaves. The white flowers are arranged in umbels and each umbel is borne on a long peduncle and has 15-25 rays that are about one inch or more in length. The fruits are brown oblong-ovoid, laterally flattened.

Ecology & Ethnobotany The common name of this plant is in reference to a former use of the plant in the treatment of gout. It is also commonly called Bishop's weed in reference to the purported resemblance of the leaflet shape to that of a bishop's miter.

Bishop's goutweed is widely planted in gardens and can be difficult to remove after it is established. It spreads primarily or completely by vegetative means, and it can be aggressive and crowd out native species.

WATER HEMLOCK
Cicuta douglasii   APIACEAE

      General Description. This species is found in marshes and along the edges of streams and ponds. Water hemlock is a stout perennial from fleshy, fascicled roots. Leaves are 1-2 times pinnately divided into narrowly lance-shaped, sharply toothed leaflets. The <u>veins of the leaflets terminate at the notches</u> <u>between the teeth</u>. Numerous white to greenish flowers are arranged in compound umbels. Fruits are slightly flattened with thickened ribs on the faces. The bruised foliage produces

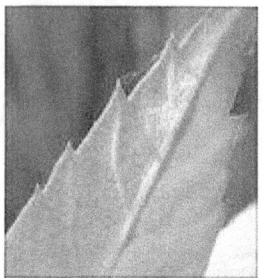

a musky odor.

      Ecology & Ethnobotany This is an ***extremely poisonous plant if ingested***. In fact, water hemlock has been described as the <u>***most violently poisonous vascular plant in North America.***</u> The whole plant contains cicutoxin, a resin-like substance that depresses the respiratory system, with the root being particularly dangerous. A single mouthful of the plant is capable of killing an adult. Water hemlocks have been used throughout the ages to execute criminals and kings. Many children have been fatally poisoned by blowing into whistles made from hollow stems of water hemlock. It is imperative that you take extra special care in your ability to identify members of the carrot family (Apiaceae). The following should reinforce this fact:

"When about the end of March 1670, the cattle were being led from the village to water at the spring, in treading the river banks they exposed the roots of this Cicuta (water hemlock) whose stems and leaf buds were now coming forth. At that time two boys and six girls, a little before noon, ran out to the spring and the meadow through which the river flows, and seeing a root and thinking that was a golden parsnip, not through the bidding of any evil appetite, but at the behest of wayward frolicsomeness, ate greedily of it, and certain of the girls among them commended the root to others for its sweetness and pleasantness, wherefore the boys, especially, ate quite abundantly of it and joyfully hastened home; and one of the girls tearfully complained to her mother she had been supplied meagerly by her comrades, with the root.

Jacob Maeder, a boy of six years, possessed of white locks, and delicate though active, returned home happy and smiling, as if things had gone well. A little while afterwards he complained of pain in his abdomen, and, scarcely uttering a word, fell prostrate to the ground, and urinated with great violence to the height of a man. Presently he was a terrible sight to see, being seized with convulsions, with the loss of all his senses. His mouth was shut most tightly so that it could not be opened by any means. He grated his teeth; he twisted his eyes about strangely and blood flowed from his ears. In the region of his abdomen a certain swollen body of the size of a man's fist struck the hand of the afflicted father with the greatest force, particularly in the neighborhood of the ensiform cartilage. He frequently hiccupped; at times he seemed to be about to vomit, but he could force nothing from his mouth, which was most tightly closed. He tossed his limbs about marvelously and twisted them;

frequently his head was drawn backward and his whole back was curved in the form of a bow, so that a small child could have crept beneath him in the space between his back and the bed without touching him. When the convulsions ceased momentarily, he implored the assistance of his mother. Presently, when they returned with equal violence, he could not be aroused by no pinching, by no talking, or by no other means, until his strength failed and he grew pale; and when a hand was placed on his breast he breathed his last. These symptoms continued scarcely beyond a half hour. After his death, his abdomen and face swelled without lividness except that a little was noticeable about the eyes. From the mouth of the corpse even to the hour of his burial green froth flowed very abundantly, and although it was wiped away frequently by his grieving father, nevertheless new froth soon took its place."

POISON HEMLOCK
Conium maculatum   **APIACEAE**

General Description Poison hemlock is a biennial with a stout taproot and a disagreeable odor when crushed. The stems are purple-blotched and hollow, and the leaves are pinnately dissected with a lacy appearance to them. The flowers are white in compound umbels, and the fruits are egg-shaped, flattened with prominent, wavy ribs. The plant is usually found in disturbed sites and waste places in the Boise Foothills. Blooms April to September.

Ecology & Ethnobotany This is an *extremely poisonous plant*. Death is said to result from the ingestion of the leaves, roots or seeds. The most famous use of poison hemlock was by the ancient Greeks as a humane method of capital punishment. It is said to be quite painless and the recipient's mind remains clear to the end. Introduced from Europe, poison hemlock has established itself as a common weed.

QUEEN ANNE'S LACE
Daucus carota   **APIACEAE**

General Description Queen Anne's lace is found roadsides, in fields, pastures, waste places, and moist clearings. This is an introduced plant from Europe, and it is the ancestor of the cultivated carrot.

Food and Survival Value The first year's roots of Queen Anne's lace can be prepared like garden carrots. We found the older roots tough and stringy. The roots can also be dried and roasted and then ground for use as a coffee substitute. The plant was used extensively by many Native Americans and should be kept in mind as an emergency food.

COMMON COWPARSNIP
Heracleum maximum   **APIACEAE**

General Description This is a stout perennial growing up 7 feet tall. The lower leaves are three lobed, resembling a maple leaf up to 14 inches long. The white flowers occur in compound umbels and the fruits are egg-shaped with only the marginal ribs winged. Cowparsnip is found in moist soils around streams, seeps, in the higher elevations such as around Mores Mountains and Boise Peak. It blooms April to June.

Ecology & Ethnobotany The young stems of cowparsnip can be peeled and eaten raw but are best when cooked. The hollow base of the plant can be cut into short lengths and used as a substitute for salt by eating or cooking with other foods. The young leaves are also edible after cooking, but we find them not very tasty. The leaves can also be dried and burned, and the ashes used as a salt substitute. Strong and bitter tasting, the cooked roots are said to be good for digestion, as well as for relieving gas, and cramps. In our experience, some plants are much more palatable than others.

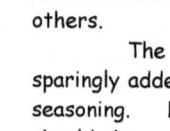

The seeds can be sparingly added to salads for seasoning. However, you should be aware that the mature, green seeds have a mild anesthetic action on tissues in the mouth. When gently chewed and sucked, they will numb the tongue and gums in a manner similar to clove oil.

The leaves of cowparsnip are large enough to be used as a toilet paper substitute and as a mild insect repellent. However, since furanocoumarin is present in the sap and the outer hairs, it may be a problem for people with sensitive skin. When the sap comes in contact with the skin in sensitive people it causes a type of "sunburn" effect (i.e., redness, blistering, and running sores) when exposed to light.

The older stems, before the flower cluster unfolds, can be peeled and the inner tissue eaten raw or cooked. While it is edible, it does have an unpleasant taste. Cooking it in a couple changes of water usually helps the taste and digestibility. In any case, cowparsnip is considered to be an excellent survival plant in the mountains. **Warning** Do not confuse this plant with other species in the same family that are highly toxic (i.e., *Cicuta* and *Conium maculatum*)

WILD PARSLEY, BISCUITROOT
Lomatium   APIACEAE

      General Description These are perennial plants with thick roots and leaves that are divided several times from the base. The white, yellow, pink, or purplish flowers are in compound umbels. The fruits are flattened and elliptical to oval in shape, and the margins may or may not be winged. Several species of biscuitroot may be encountered in the Boise Foothills. Look for them in dry ground or rocky situations.

      Ecology & Ethnobotany All species have edible roots and were an important staple among many Native Americans. They can be eaten raw or cooked, or dried and ground into flour. The flour can then be kneaded into dough, flattened into cakes, and dried in the sun or baked. Some of the species we have tried were too resinous to enjoy. Personal taste will guide one to choose the more palatable species.

      The green stems can be eaten after boiling in the springtime, but as summer progresses they become tough and fibrous. A tea can be brewed from leaves, stems, and flowers. The tiny seeds nutritious raw or roasted, can be dried and ground into meal. The plants are rich in Vitamin C.

GREAT BASIN INDIAN POTATO
Orogenia linearifolia   **APIACEAE**

General Description This is a small perennial with fleshy roots. Flowers are white and in compound umbels. Look for the species soon after snows melt in the mountains in spring and early summer. They are sometimes called Snow Drops. The species may be elusive as it is normally found early in the spring and growing in wet soils recently created by receding snows.

Ecology & Ethnobotany The roots can be boiled, steamed, roasted or baked in any way used for preparing potatoes. When small, they can be eaten raw. They can also be cooked and mashed into cakes for drying, and when protected from moisture, will keep a long time. The hard cakes can be soaked and cooked in stews.

SWEETCICELY, SWEETROOT
Osmorhiza   APIACEAE

General Description These are herbaceous perennials from stout roots, with leaves twice divided into 3's. The flowers are borne in open, compound umbels that arise from leaf axils. The fruit is spindle-shaped and compressed along the sides. In the upper forested portions of the Boise Foothills look for members of this genus on moist slopes and open areas.

Ecology & Ethnobotany The roots and fruits of any species should be tried for food. The roots in particular are recorded as being heavy with a sweet licorice or anise-like flavor, and where this is too strong, the roots and seeds can be dried and pulverized for use as a food flavoring.

YAMPAH
Perideridia   APIACEAE

General Description Yampah are biennial or perennial herbs with fascicled tuberous roots and pinnate leaves. The calyx-teeth are well-developed. The petals are white or pinkish, the stylopodium conic. The fruit is nearly terete or somewhat flattened laterally.

Comparison of *Perideridia* species

Bolander's yampah *(P. bolanderi)* Main leaves are somewhat dissected; petioles dilated; fruit oblong in shape

Gardner's Yampah (*P. gardneri*) Main leaves are only once or maybe twice pinnate or ternate; petioles not dilated; fruit oval in shape

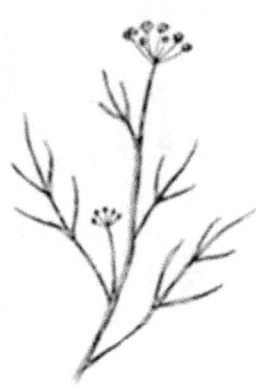

Ecology & Ethnobotany All of the species within this genus are edible. They were an important food of many indigenous peoples from British Columbia to California and the Great Basin region. The raw finger-like roots have a pleasant, nutty flavor when eaten raw, and resemble carrots when cooked. They are best when dug up before the flowers appear. The roots should be washed and peeled before cooking. They can be easily dried and will keep well for future use. When dried, the roots can be pounded and ground into flour or mashed into cakes. The seeds may be parched and ground or eaten whole.

DOGBANE
Apocynum   APOCYNACEAE

General Description Two species may be encountered in the Boise Foothills. These are perennial herbs with milky juice, that have leaves that are opposite, and the pink, bell-shaped flowers are borne in cymes.

Comparison of *Apocynum* species

Spreading dogbane (*A. androsaemifolium*) Flowers pinkish in color
Indian hemp (*A. cannabinum*) Flowers greenish in color

Ecology & Ethnobotany Dogbanes should be considered poisonous to humans if ingested. The primary use of dogbanes is for fiber. The stem fibers are strong and can be used for rope making, mats, baskets, bowstrings, fishing lines and nets, sewing, animal trap triggers, snares, cordage for bow and drill fire making, and general weaving. One of the easiest ways to isolate the fibers is to soak the stems in water.

MILKWEED
Asclepias   APOCYNACEAE

General Description A few species of milkweed could be encountered in the Boise Foothills. In general, they are erect or decumbent herbs from deep perennial roots. The leaves are opposite or whorled and the corolla is deeply 5-parted with the segments reflexed. The corona hoods each have an incurved horn within. The larvae of the monarch butterfly (*Danaus plexippus*) feeds on the leaves of milkweeds.

Food and Survival Value These are a group of important fiber plants for hikers and campers. The strong fiber can be obtained from the inner bark to make rope, fishing line, clothing, and nets. Archeologists have

discovered clothing that was made from the fibers more than 10,000 years ago. The silky floss found in mature milkweed seed pods were used in making candlewicks, and the fiber can be spun like cotton. The floss is buoyant and water resistant and makes a good insulator. The dried pods were used as utensils. The sap was used as an adhesive.

YARROW
Achillea millefolium   ASTERACEAE

General Description This is a strongly scented perennial herb with alternate leaves that are finely dissected and appear feathery. The white or sometimes yellow flowers are borne in a flat-topped corymb. Yarrow is widespread and can be found in a variety of habitats from low elevations to above timberline.

Ecology & Ethnobotany Yarrow is often referred to as "poor man's pepper." The leaves can be dried, ground, and used as seasoning. The young leaves can be added to salads. The aromatic leaves were also placed in freshly split fish to expedite drying. Rubbing the plant on one's clothing and skin, was an ancient prescription for repelling biting insects. The stalks burned on coals were said to deter mosquitoes. The leaves were used in herbal snuffs and smoking tobaccos. Yarrow has  also been used as a hops substitute for brewing yarrow beer.

MOUNTAIN DANDELION
Agoseris   ASTERACEAE

General Description These are annual or perennial, tap-rooted herbs with milky juice that resemble the common dandelion (*Taraxacum*). The flowers are all ray, yellow or occasionally orange in color.

The pappus is white with hair-like bristles. The fruit (achene) is conspicuously ten-nerved. Mountain dandelions occur on moist to dry ground, in meadows and open areas at various elevations within the Boise Foothills.

Ecology & Ethnobotany
The leaves and roots of some species are edible when cooked but are bitter, especially in late season. The seeds were eaten by the Native Americans. The sap from the leaves of some species, when hardened can be used as chewing gum. Since the sap from some species is very thick and insoluble, it may be useful for waterproofing containers (e.g., coiled baskets) and footwear.

RAGWEED
Ambrosia   ASTERACEAE

General Description The genus *Ambrosia* contains annual or perennial herbs and shrubs with pubescent and frequently glandular herbage, sometimes strong smelling. In the past, many of the shrubby species were included in the genus *Franseria*. The various species are considered to be monoecious with separate male (staminate) and female (pistillate) flowers occurring on the same plant; usually with the male flowers arranged in spikelike nodding clusters at the top of the plant for better pollen dispersal. The female flowers are arranged below the male flowers and develop into a hard bur that has hooks and spines for distributing the seeds within.

Ecology & Ethnobotany Since the wind-blown pollen is highly allergenic, ragweeds are notorious for causing hayfever where the plants are common. Ambrosia is from the Greek meaning "food of the Gods" and is the classical name for various plants. Its application to these weedy specimens is obscure.

WESTERN PEARLY-EVERLASTING
Anaphalis margaritacea   ASTERACEAE

General Description This is a rhizomatous perennial with distinctive white, woolly leaves and stems. The flowering heads are composed of disk flowers with yellow flowers surrounded by conspicuous, papery white involucral bracts. The pappus is comprised of capillary bristles. Pearly-everlasting can be found in various habitats in the upper portion of the Boise Foothills.

Ecology & Ethnobotany The herbage of western pearly-everlasting has been used as a tobacco substitute to relieve. As a tea, the plant has been used for colds, bronchial coughs, and throat infections. The whole plant can be used as a wash or poultice for external wounds. It has also been used for rheumatism, burns, sores, bruises, and swellings.

EVERLASTING, PUSSYTOES
Antennaria dimorpha ASTERACEAE

General Description These are herbaceous often mat-forming perennials. The heads are discoid, with small, white flowers surrounded by bracts that are

typically hairy below with a smooth and membranous portion varying in color from white to pink to dark brown or black. The pappus is composed of numerous hairy bristles. The various species may be encountered along in dry, open habitats, or in moist or seasonally wet places in the Boise Foothills and higher.

Ecology & Ethnobotany The sap from the stem of most species can be chewed like gum and has some nutritive value. Leaves can be used as a poultice for use on bruises, sprains, and swelling. The blossoms could be boiled and used to bathe sore or ulcerated feet or mashed and applied to sores.

STINKING CHAMOMILE
Anthemis cotula ASTERACEAE

General Description Also known as Dog Fennel, this often profusely branched annual with ill- smelling herbage grows up to 20 inches tall. The leaves are alternate and 2-3 times pinnately dissected into countless very narrw segments. Flowering heads have white rays and resemble a daisy. The bracts of the involucre are narrow and taper to a long point. In the

Boise Foothills, this species grows along roads and in other disturbed areas.

Ecology & Ethnobotany Stinking chamomile is native to Eurasia and introduced throughout North America. It is considered a noxious weed in some states. As the name suggests, its leaves emit an unpleasant odor. The plant is allelopathic, releasing chemicals that inhibit the growth of competing plants allowing it to form dense colonies in places.

COMMON BURDOCK
Arctium minus   ASTERACEAE

General Description Common burdock is one of two species that may be encountered in the Boise Foothills. Introduced from Europe, it is a coarse biennial that grows up to five feet tall. The leaves are heart-shaped, with the lower surface slightly hairy, the upper surface glabrous. The heads are discoid with purple flowers. The narrow, hook-tipped involucral bracts spread when in fruit to form the familiar "sticky" burs. Common burdock is a familiar weed of waste places.

Ecology & Ethnobotany Rich in vitamins and iron, the young leaves and shoots can be gathered for use as a potherb or eaten raw in salad. The plant has a strong rank taste and an objectionable odor. The inner pith-like material of the young stems can be eaten raw, but we find it better when boiled in one or two changes of water. The roots of young plants can be sliced and cooked, then eaten. The older roots can be roasted and ground for use as a tea or coffee substitute. Seeds can be dampened and grown as sprouts.

The tall rigid stems can be used as drills for primitive fire-starting techniques. The burs can be used as a survival "velcro" for holding clothes together.

HEARTLEAF ARNICA
Arnica cordifolia   ASTERACEAE

General Description This plant grows from 8-16 inches tall and has the lower stem leaves larger than the middle or upper ones. The leaves are generally long-petioled with heart-shaped blades. The flowering heads

are large and cup- or bell-shaped, 1 or occasionally 3 on each stem. The bracts of the involucre evidently hairy and the pappus is white. Clusters of sterile, basal leaves are common, particularly in deeply shaded areas. This plant can be found in the higher elevation above the Boise Foothills in the vicinity of Mores Mountain.

Ecology & Ethnobotany All the species are reported to be poisonous if taken internally. arnica contains arnicin, choline, a volatile oil, arnidendiol, angelic and formic acid, and other unidentified substances that can alter cardiovascular activity. The Federal Drug Administration considers arnica as "unsafe" and bans its use for human consumption. The chopped plant is steeped in rubbing alcohol for about a week and squeezed through a cloth. The liniment is then used for joint inflammations, sprains, and sore muscles. It should not be used if the skin is broken since it is toxic if it enters the bloodstream. Wear gloves as the volatile oils can be absorbed. **Warning** All species of arnica are reported to be poisonous if taken internally.

GREAT BASIN SAGEBRUSH
Artemisia tridentata   ASTERACEAE

General Description This is an evergreen shrub grows 1 to 9 feet tall and has silvery gray herbage. The leaves are narrow and spatulate, 3/8 to 1½ inches long, and with a 3-toothed apex. The inflorescence is 6 to 16 inches long, and composed of many, small, greenish flower heads. The involucres are 1/8-inch long and there are no ray flowers. There are 3 to 16 disk flowers per head, and there is no pappus. Basin sagebrush is quite common on dry slopes in the Boise Foothills. Blooms from August to October.

Food and Survival Value
The seeds are edible raw or as flour. Since many species are aromatic, they can be used to store buried food caches by masking the odor of foodstuff, and to rub on the body to mask human scent while hunting. The wood of *A. tridentata* is a good material for fire drills. Although cordage can be made from the bark, it is not very strong.

MUGWORT, WORMWOOD
Artemisia   ASTERACEAE

General Description There are a number of species of *Artemisia*, including annual, biennial, and perennial herbs and shrubs. They are mostly aromatic with entire or dissected leaves. The flower heads are small, inconspicuous, and comprised of disk flowers. The following three species are commonly encountered in the Boise Foothills.

Mugwort (*A. douglasiana*) is a perennial growing 20 to 60 inches tall, and has leaves, that are 2 to 6 inches long, lanceolate to elliptic in shape, and entire or few-toothed; green or slightly woolly above, gray woolly beneath. The inflorescence is leafy and elongate with 1/8-inch high, greenish involucres that are mostly covered with wool. There are 6 to 10 ray flowers, but they are inconspicuous. There are also 10 to 25 disk flowers. Mugwort grows in low places up to 6,000 feet. Flowers from June to October.

Tarragon (*A. dracunculus*) is an erect perennial plant that grows 20 to 60 inches tall. The leaves are linear, 1½ to 3 inches long, entire or cleft into linear lobes. The flowering heads are small, greenish, and spreading or nodding. The involucres are up to 1/8 inch across, and the flowers are all disk, with 20 to 30 per head. Tarragon occurs on dry, disturbed places and blooms from August to October.

Western mugwort (*A. ludoviciana*) is a perennial herb that grows 1 to 3 feet tall and has stems that are slightly white woolly above. The leaves are linear-lanceolate to elliptic in shape, 1 to 4 inches long, and entire or with a few teeth, white woolly on both surfaces or glabrous above. Inflorescence is elongate, and the involucres are about 1/8-inch high and is covered with wool. There are 5 to 12 ray flowers, but they are inconspicuous. There are 6 to 20 disk flowers. Western mugwort grows on dry, open places below 8,500 feet. Flowers from July to September.

Food and Survival Value The seeds of many *Artemisia* species are edible raw or as flour. The seeds and peeled shoots of *A. douglasiana* and *A. ludoviciana* were eaten raw by Native Americans.

ARROWLEAF BALSAMROOT
Balsamorhiza sagittata   ASTERACEAE

General Description These are low perennial herbs with thick rootstalks, and the leaves are mostly basal, large, and long-petioled. The yellow flowering heads are large showy, mostly on long peduncles. Balsamroot is often confused with *Wyethia* (mule's ears), which can be found in similar habitats. However, *Wyethia* leaves lack the fuzzy gray appearance seen on the balsamroot. Arrow-leaved balsamroot is commonly encountered in the Boise Foothills.

Ecology & Ethnobotany Although arrow-leaved balsamroot is considered one of the most versatile sources of food, it is not necessarily palatable. The plants contain a bitter, strongly pine-scented sap. The large taproot, root crowns, young shoots, young leafstalks and leaves, flower bud stalks, and the seeds were all eaten by various Native Americans. The larger mature leaves were often used in food preparation (i.e., wrapping).

The woody taproot of perhaps all species is edible raw or cooked. The polysaccharide inulin is the major carbohydrate found within the root. The roots can be collected throughout the year but are very difficult to dig out. In some species, the taproot may be as large as one's forearm. Cooking the roots is yet another challenge. One method we have used involves peeling the roots by pounding them to remove the bark. These were then pit cooked for 24 or more hours. When properly cooked, the roots turn brownish and sweet tasting. Another way to prepare the roots is to pit steam large quantities for a day and then mash and shape them into

cakes for storage. Cooked this way, the roots were called "pash" or "kayoum."

The young shoots are edible raw or pit cooked before they emerge in early spring. The young stems and leaves can also be eaten raw or boiled as greens. The older stems are fibrous, tough, and will require some additional boiling.

The flower bud stalks are collected while the buds are still tightly closed, then peeled and eaten raw or cooked as a green vegetable. They have a slightly nutty taste.

When harvested from dried heads, the seeds can be roasted and eaten or ground into flour. The chaff is usually removed by winnowing.

NODDING PLUMELESS THISTLE
Carduus nutans   ASTERACEAE

General Description This is a biennial with a strong, simple or sparingly branched stem that grows up to 7 feet tall. The stems are winged and the wings beset the spines. Leaves are deeply lobed to pinnately divided and the lobes are tipped with a spine. Smaller spines are distributed along the margins. The purple flower heads are solitary, large, and nodding.

Ecology & Ethnobotany *Carduus* is distinguished from *Cirsium* in that the pappus of *Carduus* is simple and smooth, not a plume. These are weedy species that may occasionally be found along roadsides and other waste places at the lower elevations.

KNAPWEED, STARTHISTLE
Centaurea   ASTERACEAE

General Description The species in this genus are annual, biennial, or perennial herbs. The herbage is often densely white-hairy when young, becoming glabrous with age. The flowering heads terminate the branches of open, leafy inflorescences. Flower heads are comprised of disk flowers, but the marginal flowers are often sterile and have an enlarged corolla making it appear as a ray flower. Flowers are white to various shades of purple and blue.

Ecology & Ethnobotany Historically, knapweed was used as a topical vulnerary, sore throat remedy, and an appetite stimulant. *Centaurea cyanus* is considered a powerful nervine, and Native Americans used it for venomous bites, indigestion, jaundice, and eye disorders. Culpeper writes that "Knapweed gently heals up running sores, both cancerous and fistulous, and will do the same for scabs of the head." Though the formulations and preparations used might be considered "questionable", the plants are abundantly available and probably warrant further investigation.

One species, *C. maculosa*, has become a serious weed in many areas. In some areas, it grows so profusely that it crowds out other species of plants, making the area uninhabitable for native plant and animal species. This genus apparently causes an inability to swallow if ingested by horses, resulting in death.

DOUGLAS' DUSTYMAIDEN
Chaenactis douglasii   ASTERACEAE

General Description These wildflowers are native to western North America, especially the southwest desert of the United States. They are quite variable in appearance. They are generally aster-like in appearance with many disc florets in each head. There may be only disc florets, but sometimes there are also enlarged ray florets along the edges. They may be white to yellow or pink.

Ecology & Ethnobotany Douglas' dustymaiden (*C. douglasii*) was first described for science in 1840 by Sir William J. Hooker (1785-1865), who dedicated the specific epithet to honor the Scottish botanist David Douglas

(1798-1834). Douglas discovered hundreds of new plants during his explorations of the American west.

An infusion of the *C. douglasii* was used as a wash for chapped hands, insect bites, boils, tumors, and swellings by the Okanagon, and Thompson. A decoction of the plant was used for indigestion, coughs, and colds. A strong decoction of the plants was applied to snakebites by the Thompson, Okanagon, and Paiute.

RUSH SKELETONWEED
Chondrilla juncea   ASTERACEAE

General Description This is a perennial with many branched, wiry stems that range from 1 to 4 feet tall. The rosette leaves resemble common dandelion and are hairless with deep, irregular teeth that point back toward the leaf base; they wither by flowering time. The plant has milky juice; coarse, reddish downward-pointing hairs at the base of the single flowering stem; and small yellow flowers and plumed seeds that ride the wind.

Ecology & Ethnobotany This plant thrives in well-drained sandy or gravelly soils and has invaded extensive areas of shallow silt loam soils in other areas as well. In addition to deep (8+ feet) taproots, it has lateral roots that produce daughter rosettes. Plants also grow from buds on root fragments cut by cultivation or other equipment.

This plant is considered a very troublesome weed in many areas. It easily invades fields, clogs harvesting machines, and successfully competes with other plants for water.

Rush skeletonweed is an aggressive wind dispersed colonizer that also spreads by creeping roots. It has been documented to invade cheatgrass dominated areas in Idaho and sagebrush communities without disturbance. Rush skeletonweed invades dry rangelands and will potentially displace native species while minimizing forage for livestock and wildlife.

GREEN RABBITBRUSH
Chrysothamnus viscidiflorus   ASTERACEAE

General Description This is an erect, much-branched shrub. The leaves are linear to lance-shaped and flat to twisted. Flowers are aggregated into heads in rounded clusters at the stem ends. Petals are yellow. This

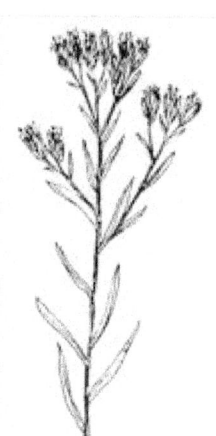

shrub is common in dry, open areas in the Boise Foothills and flowers from July to September.

Ecology & Ethnobotany Rabbitbrush produces a high-quality rubber called chrysil that vulcanizes easily. Because of the rubber-based compound, rabbitbrush will burn even if it is wet or green. Navajo Indians extracted a yellow dye from the flowers, while the inner bark yielded a green dye.

CHICORY

Cichorium intybus   ASTERACEAE

General Description This is a perennial herb that grows up to three feet tall with dandelion-like leaves. The blue flower heads, which can be seen from spring to fall, are composed of 15-20 or more ray flowers. The sap is milky. Chicory is a plant of waste places in the Boise Foothills.

Ecology & Ethnobotany While the roasted root was used as coffee, though it is not considered a very satisfactory substitute by itself (bitter but no caffeine buzz). Many coffee producers have used chicory as a coffee additive.

The young basal leaves and flowers buds hidden at the base of the leaves are edible and best if collected from fall to spring. Because they are bitter, we found it necessary to boil them in at least 1 to 3 changes of water. When collected very young, the plants are milder when eaten raw.

THISTLE

Cirsium   ASTERACEAE

General Description Thistles are characterized as biennial or perennial herbs with alternate leaves that

are lobed or cleft with spines. The red, yellow, or white heads are showy and the involucral bracts are overlapping. The native and introduced species can be found in a wide variety of habitats in the Boise Foothills into the higher elevations. At least 5 species of *Cirsium* have been recorded in the area.

Ecology & Ethnobotany There seems to be little to the members of this genus, outside of the spines, which are not edible. Flowers, seeds, young leaves, and the inner parts of the stems are all edible. The roots, crown, and inner stems can all be cooked for food. Roasting also reduces a bitter quality to the seeds.

Thistles were not a major food source in the past but were used when needed. Here is our favorite story about how useful thistles can be in emergency situations. Truman Everts, a participant in the early explorations of the Yellowstone Park region, became lost for more than a month and subsisted on thistles. He apparently had lost his glasses and was able to identify thistles by touch.

When well-dried and de-thorned, the stems can be used as hand drills for starting fires. The stem fibers of any thistle species can be used as thread or crude cordage. To obtain the fiber, simply soak the stalks in water for a day or more to loosen them from the outer layer. The downy part of seed heads makes good insulating material and a good tinder additive.

CANADA HORSEWEED
Conyza canadensis　ASTERACEAE

General Description Also known as *Erigeron canadensis*. This is an annual weed which grows to about two feet tall with numerous, narrow leaves. There are numerous white flower heads. Canada horseweed is usually found growing in waste places in the Boise Foothills.

Ecology & Ethnobotany A native to North America, horseweed was introduced into Europe around the mid-17th century where it became widely known for its tonic and astringent properties. A tea was made from the entire dried plant and used for gravel dropsy, diarrhea, and scalding urine. The leaves and tops of horseweed can be pounded and eaten uncooked. Native Americans used the plant in the form of a tea for leucorrhea and applied the solution to external sores in cases of gonorrhea.

GOLDEN TICKSEED
Coreopsis tinctoria　ASTERACEAE

General Description This is an annual that is native to the southern United States and has spread throughout much of North America. The leaves are finely divided and occurring mostly in the lower portion of the plant. The flowers are a vibrant yellow with maroon centers. The genus name comes from the Greek words *koris* meaning bug and *opsis* meaning like in reference to the shape of the seed which resembles a bug or tick.

Ecology & Ethnobotany The specific epithet means "used in dyeing." Plants in the genus *Coreopsis* are commonly called tickseed in reference to the resemblance of the seeds to ticks.

The Zuni people use the blossoms of the *tinctoria* variety to make a mahogany red dye for yarn. This variety was formerly used to make a hot beverage until the introduction of coffee by traders. Women also use an infusion of whole plant of this variety, except for the root if they desire female babies

HAWKSBEARD
Crepis acuminata   **ASTERACEAE**

General Description In general, these are perennial, tap-rooted herbs with milky juice. The leaves are alternate or all basal, and the yellow flowers are all ray. The various species can be found in dry open places This is a rather difficult genus to work with in the field as the hybridization, apomixis (asexual seed production), and polyploidy (multiplication of entire chromosome compliments) are common.

Ecology & Ethnobotany The stems and leaves of *Crepis* were eaten by Native Americans. The Karok Indians of Northern California peeled the stems of *C. acuminata* before eating. The seeds or whole plant *of C. acuminata* was thoroughly crushed and applied as a poultice to breasts after childbirth to induce milk flow. The root of the plant was used to remove a foreign object from the eye. The root can also be ground into a smooth powder and sprinkled in the eye to treat eye problems. Several applications were necessary.

RABBITBRUSH, GOLDENBUSH
Ericameria   ASTERACEAE

General Description The species here were previously included in other genera such as *Chrysothamnus* and *Haplopappus*. These are herbs or shrubs that are often glandular. The leaves are alternate, entire to pinnatifid, and often thick. There is another rabbitbrush considered separately in this handbook and is known as green rabbitbrush (*Chrysothamnus viscidiflorus*).

Ecology & Ethnobotany The seeds and stems of some species were eaten by Native Americans. A decoction from the stem of *E. cuneatus* was used to treat colds.

A tea was reported to be made from the twigs of rubber rabbitbrush (*E. nauseosa*) that provided relief from chest pains, coughs, and toothaches. The leaves and stems were also boiled and the liquid was used to wash itchy areas.

Great Basin Indians were accustomed to chewing the stems of rubber rabbitbrush to extract the latex. They believed that chewing rabbitbrush relieved both hunger and thirst. The secretion obtained from the top of the roots can also be chewed as gum.

FLEABANE
Erigeron   ASTERACEAE

General Description There are at least 5 species of fleabane in the area. They are characterized as annual, biennial, or perennial herbs with alternate or basal leaves. The flowering heads are radiate with narrow ray flowers that may be white, pink, blue, purple, or occasionally yellow. The numerous disk flowers are yellow, and the pappus is comprised of capillary bristles.

Ecology & Ethnobotany The various species bloom mostly in the spring and early summer, except at the higher elevations, and can be found in a variety of habitats. The genus name comes from the Greek *eri* (early) and *geron* (old man). The common name, fleabane, comes from the belief that these plants repelled fleas. This genus is rather difficult to identify to species in the field.

WOOLLY SUNFLOWER
Eriophyllum lanatum   ASTERACEAE

General Description The yellow flowers are about 2 inches in diameter and consist of 5 to 13 ray florets (occasionally zero) and many tiny, glandular-hairy disc florets, which when mature form a semi-sphere at the plant center. The phyllaries also number from 5 to 13; they are all the same length, and covered with flattened white, woolly hairs that tend to obscure the divisions, at least towards the base. This is a rather common plant in the Boise Foothills.

Ecology & Ethnobotany The seeds can be gathered, parched and then ground into flour.

ENGELMANN'S ASTER
Eucephalus engelmannii   ASTERACEAE

General Description This perennial herb has simple alternate and entire leaves. The outer phyllaries are strongly veined and shingled; inner phyllaries are purple-tinged. Some outer phyllaries are quite long, strongly veined, and sharply pointed. The few ray flowers (10-17) range from white to pink, especially turning pink as they age. Pappus are of capillary bristles.

Ecology & Ethnobotany The species was named *Aster elegans* by D. C. Eaton in 1871 and renamed *A. engelmannii* by Asa Gray. In 1896 Edward Greene renamed it to *Eucephalus engelmannii*. *Eucephalus* is Greek for well-developed head. George Engelmann was an eminent St. Louis physician and was one of the top three American botanists of the 19th century (along with John Torrey and Asa Gray).

WESTERN MARSH CUDWEED
Gnaphalium palustre   ASTERACEAE

General Description These are woolly annual, biennial, or short-lived perennials herbs with alternate leaves. The plants are often confused with plants in the genus *Antennaria*, but cudweeds have both male and female flowers on the same plant and are tap-rooted. The disk flowers are yellow or whitish. The species are found in the Boise Foothills in moist, open areas to well-drained soils.

Ecology & Ethnobotany The bruised plant assists in healing wounds and steeping the leaves in cold water is used for increasing perspiration. Some species contain pyrrolizidine alkaloids and should be regarded as potentially toxic.

CURLYCUP GUMWEED
Grindelia squarrosa   ASTERACEAE

General Description Gunweed is a taprooted biennial or short-lived perennial with several stems. Alternate, oblong, and sessile leaves are resinous and coarsely toothed to serrate or even entire. Middle and upper leaves clasp the stem. The regularly overlapping involucre bracts glisten with a gummy resin, the source of the plant's strong, pungent odor. The outer, green bracts are evidently reflexed. Ray and disc flowers are bright yellow. This plant is a common native weed of disturbed sites in the Boise Foothills.

Ecology & Ethnobotany In general, gumweeds are considered toxic, and the toxicity appears to be dependent upon the soil in which it grows. However, many species have been used medicinally for hundreds of years.

The sticky flowers heads of this species were used as a chewing gum substitute. The young leaves make an aromatic, bitter tea. The flower heads can be boiled in water and used as an external remedy for skin diseases, scabs, and sores. A hot poultice of the plant was used for swellings.

BROOM SNAKEWEED
Gutierrezia sarothrae   ASTERACEAE

General Description This is a sticky, glandular perennial, appearing almost shrubby, and grows 6 to 24 inches high. The leaves are entire, linear-filiform, and less than 1/8-inch wide. The flowering heads are small, numerous, and with whitish, leathery, involucral bracts. Ray flowers are yellow, number 3 to 8 per head, and are 1/8-inch long. There are 3 to 8 disk flowers, and the pappus consist of 2 to 8, stiff awns. In the Boise Foothills, broom snakeweed grows on dry slopes.

Ecology & Ethnobotany Another common name, matchweed, refers to the match-like appearance of the flower heads. As with many aromatic plants, this species was used medicinally. The plant was boiled to make a tea for colds, coughs and dizziness. The tops of fresh, mature snakeweed were boiled until strong and dark. The liquid could be drunk for lung trouble and colds or applied externally for skin ailments such as heat rash, poisoning, and athlete's foot. For respiratory ailments, the root was boiled in water and the steam inhaled.

SUNFLOWER
Helianthus annuus   ASTERACEAE

General Description This is a coarse annual herb with tall stems. Leaves are simple, the lower ones are opposite, others sometimes alternate. Flower heads are showy with bright yellow ray flowers. Involucral bracts are green and herbaceous.

Food and Survival Value The largest member of this genus in the area is *H. annuus*, is a valuable and useful plant. It has been cultivated in the United States since before Columbus.

The seeds may be eaten raw or roasted, then ground into meal and made into bread. The roasted shells can be used as a coffee substitute. To separate large amounts of seeds from shells, first grind them coarsely, then stir vigorously in water. In this way the shells will float, while the seeds sink to the bottom. The tiny unopened flower buds are also edible with a flavor similar to artichokes. To reduce their bitterness boil in 2-3 changes of water. Serve with lemon and melted butter.

Sunflower oil can be extracted from the seeds for cooking, and can also be used in making soap, paints, varnishes, and candles. It is extracted by simply boiling the crushed seeds and then skimming the oil from the surface of the water. The pulp remaining after the oil is extracted also provides food for livestock.

HAWKWEED
Hieracium cynoglossoides & H. gracile
ASTERACEAE

General Description Hawkweeds are fibrous rooted perennial herbs with milky juice. With one exception, the flowering heads have bright yellow to orange ray flowers, and there are no disk flowers. The phyllaries are in 1-3 series and are worth a look with a hand lens, being ornamented variously with hairs and glands. The pappus is of stiff, brown bristles. Hawkweeds are found in a variety of habitats up to the subalpine.

Comparison of *Hieracium* species

H. gracile Leaves mainly basal, glabrous or short-hairy
H. cynoglossoides Stem leaves apparent, usually with some long hairs

Ecology & Ethnobotany The name "hawkweed" comes from the belief by the ancient Greeks that hawks would tear apart a plant called the hieracion (from the Greek *hierax* meaning 'hawk') and wet their eyes with the juice to clear their eyesight. The green plant and juices of white-flowered hawkweed may be used as a substitute for chewing gum, although it is best when dried first. The plant was also used to ease toothaches, to cure warts, or as an astringent in treating hemorrhages, and as a general tonic.

POVERTYWEED
Iva axillaris   ASTERACEAE

General Description Poverty-weed is a perennial with an unpleasant odor and creeping rootstocks. The branched stems have sessile, oblong to linear, entire, thick and pale green leaves. Small, nodding flower heads are borne singly in the angle between stem and leaf. This species is found in the Boise Foothills, usually growing on disturbed sites.

Ecology & Ethnobotany This is a long-lived perennial with creeping roots. It is a widespread native that is a desirable component of salt marsh and alkali plains. In areas that have been disturbed by overgrazing, it will form large clonal colonies that once established, will be difficult to eradicate. The pollen of this plant is highly allergenic and plants may cause contact dermatitis in sensitive individuals.

PRICKLY LETTUCE
Lactuca serriola   **ASTERACEAE**

General Description This is a tall, prickly plant with alternate leaves and milky juice. The ray flowers of this species are yellow; other species may be blue, or whitish. The pappus is white to brownish. This is a rather

common weed in fields and waste places in the Boise Foothills.

Ecology & Ethnobotany Collected in the late fall to early spring, the plants should be boiled in a couple of changes of water to reduce the bitterness. The earlier or younger the plant is collected, the better the flavor. Because of the latex sap, raw greens can cause upset stomach if eaten in quantity. In sensitive people, the latex can cause dermatitis. These wild plants contain more Vitamin A than spinach and a good quantity of Vitamin C.

WHITE LAYIA
Layia glandulosa   **ASTERACEAE**

General Description This annual grows up to 24 inches tall and has leaves that are rough hairy, linear to lanceolate in shape, with the basal ones, being toothed or lobed while the upper ones are entire. The flowering heads have both ray and disk flowers present. There are about 25 to 100 disk flowers. Layia is found in sandy soil and blooms from March to June.

Ecology & Ethnobotany The seeds of this species are edible after grinding it into flour for mush or incorporated into pinole.

HOARY ASTER
Machaeranthera canescens   **ASTERACEAE**

General Description This is a taprooted, branched, and several-stemmed biennial or short-lived perennial. The leaves are alternatre, linear to inversely lance-shaped or spatula-shaped and having teeth with minute spiny tips. The inflorescence has numerous small, bract-like leaves. The numerous asterlike flowers have deep violet-purple rays and yellow to reddish disc flowers. The narrow involucre bracts have short green, reflexed or spreading tips that are occasionally suffused with red. This species is comnon on dry, open slopes of the Boise Foothills.

Ecology & Ethnobotany There are about 30 species of *Machaeranthera* in temperate North America. The generic name was derived from the Greek *machaira*, "sword" and *anthera*, "anther," in reference to the pointed, pollen-bearing organs. The Hopi Indians made a beverage from the taproot of a related species, *M. grindelioides*, which was used for coughs.

TARWEED
Madia   ASTERACEAE

General Description Typically, these are annuals with a tar scent of varying intensity. The leaves are narrow, usually opposite below and alternate above. The flower heads are comprised of inconspicuous yellow ray flowers. In the Boise Foothills, at least 3 species of tarweed can be found in open, grassy, or vernally moist areas.

Ecology & Ethnobotany The seeds of all tarweed seeds were collected and stored until needed. Seeds were often used in making pinole by Native Americans. Some tribes pulverized tarweed seeds and ate them dry. The roots of some species were also eaten.

The scalded seeds also yield a nutritious oil. The oil was used like olive oil before olive was readily available in this country.

PINEAPPLE WEED
Matricaria discoidea   ASTERACEAE

General Description This is an annual herb with a branched habit. Leaves are alternate and pinnately lobed or divided. The small, terminally arranged flower heads are composed of disk or ray flowers. This is an introduced plant with a circumboreal in distribution. It can be easily recognized by its pineapple-like smell when crush between the fingers.

Ecology & Ethnobotany A delicious tea can be made from the dried flowers of the plant. The leaves are edible, but bitter. The medicinal uses of pineapple weed are identical to that of chamomile (*Anthemis*). Used as a tea it is a carminative, antispasmodic, and mild sedative.

NODDING SILVERPUFF

Microseris nutans   ASTERACEAE

General Description This is a taprooted perennial with milky juice. The leaves are basal, and the flower heads are always ligulate and yellow. The heads are comprised of all ray flowers, and the ligules are yellow. Achenes are not beaked and the pappus is of 5, silvery, papery scales that are tipped with an awn or bristle. Look for the plant in open and moist habitats in the upper portions of the Boise Foothills.

Ecology & Ethnobotany This genus is distinguished from *Agoseris* by a rather technical character; *Microseris* has a pappus attached to the end of the narrow, elongated achene, whereas *Agoseris* have the pappus attached to the achene via a relatively long, thin beak. The slender roots of nodding microseris are apparently edible raw

SAGEBRUSH FALSE DANDELION
Nothocalais troximoides   ASTERACEAE

General Description The stem of false-dandelion rises up to 12 inches tall. Narrow leaves are less than 1/8-inch wide with wavy or crisped margins, are crowed at the base of the naked flowering stem. The lance-shaped to linear bracts generally have a dark midrib. The pappus is composed of 10-30 gradually tapering scales (not bristles).

Ecology & Ethnobotany The species was previously included in the genus *Microseris*. Sagebrush false dandelion is found in open, arid habitats and can be seen flowering during the spring and summer.

SCOTCH THISTLE
Onopordum acanthium   ASTERACEAE

General Description This is a woolly biennial with winged stems growing up to 6 feet tall. The leaves are alternate, lobed or coarsely toothed, or spiny margins. The involucral bracts are linear to lance-linear in shape. There are no ray flowers and the disk flowers are purple. The pappus is composed of barbellate bristles.

Ecology & Ethnobotany It is native to Europe and Western Asia from the Iberian Peninsula east to Kazakhstan, and north to central Scandinavia, and widely naturalized elsewhere. It is a major agricultural weed in western United States. With enough moisture, it can re-sprout from roots cut up during cultivation. This plant spreads easily because each plant can produce over 20,000 wind dispersed seeds. The seeds can also be dispersed by water or by being caught in the fur of animals.

It is grown as an ornamental plant for its bold foliage and large flowers. It has been used to treat cancers and ulcers and to diminish discharges of mucous membranes. The receptacle was eaten in earlier times like an artichoke. The cottony hairs on the stem have been occasionally collected to stuff pillows. Oil from the seeds has been used for burning and cooking.

WOOLLY GROUNDSEL
Packera cana   ASTERACEAE

General Description This plant grows up to 12 inches tall. It arises from a branched, persistent, woody base and a short taproot. The herbage is more or less white-hairy, but the upper leaf surfaces often become glabrous at maturity. The basal and lowermost stem leaves are short- or long-petiolate and mostly tufted. Leaf shape varies from normally elongated and pointed to roundish, entire, or rarely pinnately lobed. The middle and upper stem leaves are strongly reduced, approaching a bractlike condition. This is a common plant of dry exposures, including open forest.

Ecology & Ethnobotany Members of this genus were previously included in the genus *Senecio*. They are mostly perennial herbs with basal or alternate entire to deeply lobed leaves. The involucral bracts uniseriate and the ray flowers are yellow or orange or lacking. Pappus is comprised of capillary bristles. You should include vegetative basal rosette leaves (or take notes about their shape) when attempting species in this genus.

COTTONBATTING PLANT
Pseudognaphalium stramineum   **ASTERACEAE**

General Description These are somewhat woolly annual to perennial herbs. Leaves are alternate and entire. The flowering heads occur in a terminal inflorescence and there are no ray flowers. The outer flowers are pistillate (female) and the few inner flowers have wider corollas than the outer ones. The inner flowers are perfect (both sexes present) and are yellow, whitish or purplish in color. The pappus is of capillary bristles. Look for this plant in moist, open places.

Ecology & Ethnobotany The species was once included in the genus *Gnaphalium*. Other synonyms include *Gnaphalium chilense* and *G. stramineum*.

WESTERN CONEFLOWER
Rudbeckia occidentalis   **ASTERACEAE**

General Description Coneflowers are tall biennials or perennials with alternate leaves. The large flower heads have hemispheric, elongated, egg- to cone-shaped disks (central portion of the flowering head) and mostly reflexed or spreading involucre bracts. The plant is common in moist, open to partly shaded habitats in the upper forested portion of the Boise Foothills.

Ecology & Ethnobotany The genus name honors the Swedish father and son who were professors of botany and predecessors of Linnaeus - O.J. Rudbeck (1630-

1702) and *O.O.* Rudbeck (1660-1740).

RAGWORT
Senecio   ASTERACEAE

General Description These are annual, biennial, or perennial herbs with alternate or basal leaves. The flower heads are yellow and the pappus is made up of hair-like bristles. Ragworts can be found in various habitats and elevations. Three species may be found in the Boise Foothills.

Ecology & Ethnobotany *Senecio* is one of the largest genera of plants with nearly 2,000-3,000 species distributed worldwide. Approximately 100 species are found in the western United States. Many species contain highly toxic alkaloids and should therefore be avoided.

GOLDENROD
Solidago   ASTERACEAE

General Description The various species of goldenrod are perennial herbs with fibrous roots. The leaves are alternate, simple, and either tooth or entire. The heads are made up of yellow ray flowers. ALook for goldenrods in dry to moist habitats from the foothills to forested areas in the Boise Foothills. They are often growing in dense patches. Here is a dichotomous key for you to use in trying to identify species.

1a. Stems glabrous below inflorescence; leaf surfaces glabrous ----- **2a**
1b. Stems hairy, at least between middle and inflorescence; leaf surface hairy ----- **Canada goldenrod** (*S. canadensis*)

2a. Largest leaves at middle of stem, elliptic or lane-elliptic, sharply acute at tip ----- **giant goldenrod** (*S. gigantea*)
2b. Largest leaves towards the base, oblanceolate, acute or obtuse at tip ----- **Missouri goldenrod** (*S. missouriensis*)

Ecology & Ethnobotany Goldenrods are attractive to bees and butterflies. Genus name comes from the Latin words *solidus* meaning whole and *ago* meaning to make in reference to the medicinal healing properties of some species plants.

Young leaves can be prepared as potherbs or added to soups. Depending on habitat, age, and personal preference, their palatability is quite variable. The dried leaves and dried, fully expanded flowers can be used to make a tea. The seeds can be used to thicken stews. Large amounts of the raw herbage should be avoided as it may be toxic.

COMMON SOW THISTLE
Sonchus oleraceus   ASTERACEAE

General Description Common sow thistle grows up to 40 inches tall. All leaves are softly prickle-margined, and the lower ones are pinnately divided with a terminal segment that is large and triangular to irregularly lobed or only toothed. Upper stem leaves are sessile and clasping with large, prominent lobes at the base. The leaves are both gradually less divided and reduced upwards. Several flower heads are arranged in a terminal, nearly flat-topped inflorescence. The pappus is bristly. In the Boise Foothills the species can be found in waste places.

Ecology & Ethnobotany The common name sow thistle is a misnomer because this genus of plants is more closely related to the genus *Lactuca* (wild lettuce). Both groups of plants have a milky latex, unlike true thistles (*Circium*).

The young plants of all species can be prepared as a potherb. As they get older they become increasingly bitter. We found that boiling them in at least two changes of water makes them a little more palatable. Since the plants have an abundance of soluble vitamins and minerals, use only a minimum amount of water and boil briefly.

DANDELION
Taraxacum officinale   ASTERACEAE

General Description Dandelions need very little introduction. All species are tap-rooted perennials with milky juice and leaves that form a dense, basal rosette. The solitary flower head is composed of bright yellow ray flowers. They are found in a variety of habitats up to the alpine zone.

Ecology & Ethnobotany Every part of dandelion is edible. The young leaves may be eaten raw or cooked like spinach. The older leaves are also edible, but we find it is better to boil the older leaves in 1 or 2 changes of water to eliminate the bitterness that comes with age. The plants are high in Vitamins A and C, a good source of B complex, and iron, calcium, phosphorous, and potassium. The roots can also be eaten raw, or boiled as a vegetable, baked as potatoes, or added to soups and stews. The roasted root can be used as a substitute for coffee, but it lacks the caffeine buzz. The flower buds can be pickled and added to meals such as omelets.

SALSIFY, GOAT'S BEARD
Tragopogon　　ASTERACEAE

General Description The two species in the Boise Foothills are introduced, tap-rooted, biennial herbs with milky juice. The leaves are alternate, entire, sessile and clasping at the base and taper to a long point. The flower heads are solitary and composed of pale yellow flowers. The heads open early in the day, close about noon and remain closed on cloudy, rainy days. They are found in many habitats

Yellow salsify (*T. dubius*) Stem below the flowering head swollen and hollow 1
Meadow salsify (*T. pratensis*) Stem not enlarged below the flowering head

     Ecology & Ethnobotany *Tragos* is Greek for goat and *pogon* for beard, thus giving another common name, Goat's Beard. The genus was named by Linnaeus in 1753 and this species was first collected near the Adriatic Sea.

     The fleshy roots of both species can be eaten raw or after cooking. The flavor resembles that of an oyster, an acquired taste! They are somewhat fibrous and tough. Salsify root has been cultivated for over 2,000 years in the Mediterranean. The young leaves and stems of all species can be eaten after boiling until tender. The coagulated sap can be used as chewing gum and as a remedy for indigestion.

MULE-EARS

Wyethia   **ASTERACEAE**

     General Description This coarse, erect perennial has short stems, grows 4 to 12 inches tall, has large, ovate leaves that hide the heads of flowers. The leaves are alternate, broadly ovate to roundish, 2½ to 8 inches long and up to 3 or more inches wide. They are also

entire and petiolate. The flowering heads are hidden among the leaves. Both ray and disk flowers are present, but the ray ligules are short and are not as long as the outer involucral bracts. There are about 5 to 8 yellow ray flowers per head. The disk flowers number 12 to 20, and the pappus consist of a crown of unequal scales. In the Boise Foothills, this is a common species growing in grassy, open, or wooded areas. Flowers from May to August.

Ecology & Ethnobotany All species have leaves on the stems distinguishing them from *Balsamorhiza*, which only has leaves at the base.

The roots are edible after long cooking. Seeds are edible too and resemble sunflower (*Helianthus*) in taste. ***The leaves, if nothing else, are supposedly poisonous.***

COCKLEBUR
Xanthium   ASTERACEAE

General Description These are coarse annual weeds of uncertain origin that have a cosmopolitan distribution. The stems of the plants are simple, and the leaves are alternate. The flower heads are solitary or clustered in the leaf axils. The bur (seed) has conspicuous, slender hooked prickles. Cockleburs can be found in the lower portions of the Boise Foothills.

<u>Comparison of *Xanthium*
species</u>

Spiny cocklebur (*X. spinosum*)
Plants with stout 3-branched
spines in leaf axils
**Rough   cocklebur   (*X.*
strumarium)**   Plants   lacking
spines in leaf axils

   Ecology & Ethnobotany The flowerheads rely
on the wind, rather than insects, to cross-pollinate
individual plants. Because the burry fruits readily cling to
the fur of mammals and the clothing of humans, they are
easily transported to new areas, spreading the seeds of
this plant.
   The uses of cocklebur are primarily medicinal.
They have been used by many aboriginal people
throughout North and South America, and as herbal
medicine in China. The seeds were ground, mixed with
corn meal, made into cakes or balls, and steamed by the
poorer class of Zuni Pueblo.

BARBERRY

Mahonia aquifolium & M. repens   BERBERIDACEAE

   General Description These are shrubs with
pinnately compound, evergreen leaves. The leaflets have
spiny margins, and the yellow flowers are in 3 whorls that
are interpreted as bracts, sepals, and petals. The flowers
have six or more stamens that split open by two hinged
valves to splatter pollen over insects as they crawl by.
Use a hand lens to view the unique anthers, which open to
release pollen by a pressure-controlled, flap-like valve,

instead of splitting down the side. The fruits are blue to purple in color and have a waxy covering.

Comparison of *Mahonia* species

M. aquifolium Stems erect; leaflets 5-ll, strongly spinulose (spines)
M. repens stems more or less prostrate; leaflets 3-7, weakly spinulose

Ecology & Ethnobotany The blue berries are edible raw or can be dried for future use or added to

soups to improve flavor. The plants contain berberine, a bitter alkaloid that gives roots their distinctive yellow color and usefulness as a digestive tonic. Berberine stimulates the involuntary muscles and possesses anti-pyretic, laxative, and anti-bacterial qualities. A yellow dye can be obtained by boiling bark and roots.

ALDER
Alnus   BETULACEAE

General Description The species are small trees or shrubs with smooth, reddish or gray-brown bark. Leaves are egg-shaped and have serrate edges. The male catkins are grouped near the end of branches and drop off after pollen is shed. The female catkin is cone-like and persistent. Fruits are flattened achenes with lateral wings or just a membranous border. These plants are usually associated with riparian and wetland sites at low to mid elevations.

Ecology & Ethnobotany Since alders usually grow in the vicinity of free-flowing water, it is considered a botanical indicator of water. The edible catkins are high in protein, but generally don't taste very good. The catkins are more tolerable if they are nibbled raw, added to soups, or dried and powdered and used as a spice. The inner bark is  palatable only for a short time in the spring when it is less bitter. A patch of bark is removed from the tree and the tissue scraped off and eaten fresh or dried in cakes. Alder is valued for its hardwood and is useful for open fires as it does not readily spark. It is used widely by Native Americans for woodworking, including dishes, spoons, and platters. The wood is also used for making fire drill sets.

BIRCH
Betula occidentalis   BETULACEAE

General Description This is a deciduous shrub with simple, alternate, and sharply toothed leaves. Birches can be found along streams, and in wet meadows and bogs from the foothills to upper montane zone.

Ecology & Ethnobotany Young birch leaves can be added to salads. The inner bark can be dried and ground into flour, and the twigs can be steeped in hot water for a tea. The juice of birch leaves makes a good mouthwash.

Birch contains a significant amount of methyl salicylate and is often used in teas for headaches and rheumatic pain. Birch is highly regarded as a medicinal plant in Russia and Siberia for treating arthritis.

Since it burns even when wet, birch bark makes a good tinder. The sap, collected in much the same manner as maple, was sometimes made into syrup or vinegar. The best time for tapping is early spring, before the leaves unfurl.

FIDDLENECK
Amsinckia   BORAGINACEAE

General Description Fiddlenecks are coarse annual herbs with stiff hairs. The flowers are in a scorpion tail-like spike. The genus is named after W. Amsink, an early 19th century patron of the botanic garden in Hamburg, Germany.

Ecology & Ethnobotany The seeds of *A. menziesii* were pounded into flour, then made into cakes and eaten without cooking.

MADWORT

Asperugo procumbens   BORAGINACEAE

General Description This annual plant has weak, climbing or trailing stems that grow up to 2 feet tall. The lower leaves are petiolate and the lance- to egg-shaped blades are 1-4 inches long. The upper leaves are smaller and have no petioles. Stems and leaves are covered with fine, stiff hairs. The flowers are blue and are borne on short, curved stalks from the base of the leaves.

Ecology & Ethnobotany This is an introduced plant from Eurasia that has become widespread in the northern part of the United States.

CRYPTANTHA

Cryptantha   BORAGINACEAE

Genera Description These are annual or perennial herbs that are rough to the touch. They have linear or spatulate leaves and the inflorescence is scorpion tail-like with small, white or yellow flowers. The corollas are small and white with well-developed appendages inside below the lobes (fornices). Small-scale characteristics separate *Cryptantha* from other borages.

Ecology & Ethnobotany Cryptantha are usually found in dry, open areas at various elevations in the Boise Foothills. Characters of the nutlets such as surface texture, size, number maturing, and size and shape of attachment scar are generally consistent for each species and serve as useful diagnostic features. They are especially important in annual species, since the annuals are often vegetatively similar. Also, these characters are not always evident except in fully mature fruit, a time of phenology when the plants are seldom noticed or collected.

The genus is native to western North America and South America. The seeds of some *Cryptantha* may have been eaten by some Native Americans.

STICKSEED
Hackelia   **BORAGINACEAE**

General Description Members of this genus are taprooted biennial or perennial herbs with entire leaves. The inflorescence is composed of narrow, leafy branches that elongate in fruit. The flowers are blue or white, often with a yellow center. The appendages in the throat of the corolla are well-developed. The nutlets have prickles along the margins and sometimes over the entire surface. Two species may be encountered in the Boise Foothills: gray stickseed (*H. cinerea*) and manyflower stickseed (*H. floribunda*). The former has white flowers, the latter blue.

Ecology & Etnobotany Each of the flowers gives rise to four small nutlets that possess rows of barbed prickles down their edges - hence the name stickseed. It is these prickles that readily enter clothing or the fur of animals but retard their being pulled out again. The seeds are then transported and the plants become established a long distance from where they originated.

GROMWELL, STONESEED
Buglossoides arvensis & Lithospermum ruderale
BORAGINACEAE

General Description These are plants with erect and pubescent or hairy stems. The leaves are alternate and entire, and their veins are usually indistinct. The inflorescence is leafy-bracted and the flowers are white, yellow, or blue. The calyx is narrowly 5-parted or cleft and the corolla is 5-lobed. Stigma is capitate or 2-lobed. There are 4 erect nutlets, and in most species, they are white and shining.

Comparison of "Stoneseed" species

Buglossoides arvensis Annual; stems solitary; flowers white to cream
Lithospermum ruderale Perennial; stems several; flowers yellow

Ecology & Ethnobotany The annual *species L. arvensis* has been placed in the genus *Buglossoides*. This species has 1-several simple or branched stems 4 to 28 inches tall. Leaves are linear to narrowly lance-shaped in outline, usually without petioles, and $\frac{1}{2}$ to $2\frac{1}{2}$ inches long. The flowers are borne in the leaf axils. The inflorescence is crowded at first, becoming open with age. Flowers are white, sometimes with a bluish cast. The nutlets are wrinkled and pitted, occasionally with scattered bumps.

Western gromwell (*Lithospermum ruderale*) is a perennial herb up to 2 feet tall with a woody taproot and greenish yellow flowers. It is common in dry grassland areas and open forests up into the middle elevations. The genus name means "stone seed," referring to the hard nutlets.

    Lithospermum was used by Native Americans throughout the West as a medicine and food. Although little is known about its chemistry, the effectiveness of several species as contraceptives and as depurative for skin conditions warrants further investigation.

BLUEBELLS, LUNGWORT
Mertensia longiflora & M. oblongifolia
BORAGINACEAE

    General Description The members of this genus are perennial herbs with 1-several stems from a rhizome, caudex, or corm. The foliage is usually glabrous to sparsely hairy, but not bristly. The leaves are alternate and entire-margined. The inflorescence consists of 1-many drooping to erect clusters in the axils of leaves and terminating the branches. Flowers are blue. The petals are united almost their full length. The corolla, consisting of a narrow tube and an expanded limb, is trumpet- to bell-shaped. The appendages at the base of the limb (fornices) are usually apparent. Nutlets are generally roughened.

Comparison of *Mertensia* species

M. *longiflora* Stems 1-2 from a short tuberous root;
basal leaves absent
M. *oblongifolia* Stems several from a branched root;
basal leaves present

Ecology & Ethnobotany Bluebells are often
overlooked in many edible plant guides. The flowers can
be nibbled upon raw or added to salads. Since the leaves
are a bit hairy, we found them better when chopped up
and added to soups. Bluebells may contain alkaloids and
other constituents that can be toxic if consumed in large
quantities.

FORGET-ME-NOT
Myostis   BORAGINACEAE

General Description In general, these are annual,
biennial, or perennial herbs with alternate, entire leaves
that are strongly veined. The flowers are blue, pink, or
white, and are divided into a distinct tube and limb. The
nutlets are smooth and shiny. The genus name is from the
Greek, *mys*, meaning mouse, and *otis*, meaning ear,
referring to the hairy leaves. Two species in the Boise
Foothills: *M. laxa* and *M. stricta.*

Ecology & Ethnobotany These plants are pollinated
by insects even though the pollen is very small and is
usually indicative of wind pollinated plants.

There are a number of stories associated with the
common name of forget-me-not. The best known tells of
a couple strolling along the Danube River in Europe. While
walking, the lady noticed a clump of pretty blue flowers
on the streambank and her lover tried to pick some, but

as he reached for them, he slipped and fell into the water. As he fell, he was able to toss the flowers to the lady and call out "forget me not"!

COMBSEED
Pectocarya   BORAGINACEAE

General Description These are small annual plants which bear tiny white flowers no more than 1/8  inch in diameter. Their fruits are nutlets which often have small projections that look like the teeth of a comb, hence their common name. The nutlets usually come in clusters of four.

Ecology & Ethnobotany An early spring plant in the Boise Foothills.

PACIFIC POPCORNFLOWER
Plagiobothrys tenellus   BORAGINACEAE

General Description The species is an annual with alternate or opposite leaves. The white, salver-form flowers are in scorpioid racemes. The various species occur in moist soil.

Ecology & Ethnobotany *Plagiobothrys* and *Cryptantha* are difficult for many people to distinguish. In order to accurately identify the species of both genera, it is necessary to have both the flowers and fruits (nutlets) present.

ALYSSUM, MADWORT
Alyssum　　BRASSICACEAE

General Description These are small annual plants with foliage that appears dull gray due to the presence of dense, star-shaped hairs. The leaves are alternate and simple, and the flowers are short-stalked on the terminal portion of the stems. The flowers have light yellow petals that quickly fade to white. Fruits are egg-shaped or round in outline and have winged margins and a short style. Two species can be found in the Boise Foothills.

Ecology & Ethnobotany Both species are Eurasian weeds and are widespread in distribution throughout most of the United States. The genus name is from the Greek *a* (without) and *lyssa* (rabies), to the supposed cure for rabies. The small leaves of pale madwort are mild tasting and can be eaten raw.

ROCKCRESS
Arabis & Boechera　　BRASSICACEAE

General Description These are biennial or perennial herbs with stellate hairs. Flowers occur in racemes and are usually white to purple in color. The fruits are linear siliques, usually flattened parallel to the partition. Several species may be found in the Boise Foothills and occur in a variety of habitats.

Ecology & Ethnobotany Though *Arabis* was traditionally recognized as a large genus with many Old World and New World members, recent studies of these species using genetic data suggest that they are not closely related, so *Arabis* has been split into two separate genera. The Old World members all remain in the genus *Arabis*, whereas most of the New World members have been moved into the genus *Boechera*, with only a few remaining in *Arabis*. From an edibility perspective, they are similar.

WINTERCRESS
Barbarea orthoceras   BRASSICACEAE

General Description This is an erect glabrous perennial with stout angled stems. The basal leaves are 1-4 inches long, pinnatifid with a large terminal leaflet. The leaves on the main stem are pinnatifid or lobed. Flowers occur in a dense raceme and are pale yellow in color. The mature fruits are 1-2 inches long. In the Boise Foothills, look for wintercress in moist places and along stream banks.

Ecology & Ethnobotany The young plants are best when gathered in early spring. They are good in salads or as a potherb. Older leaves can still be eaten by parboiling them to remove the bitterness. The flowers clusters, even when opened, can be substituted for broccoli.

HOARY FALSE MADWORT
Berteroa incana   BRASSICACEAE

General Description This annual grows up to 32 inches tall that are branched both at the base and above. The lower leaves are broadly linear and entire-margined with short petioles; these wither as the plant matures. The upper leaves are smaller and sessile. The foliage is grayish with a dense covering of branched hairs. The flowers are borne on short stalks on the terminal portion of the stem. The notched petals are white and the slightly inflated fruit s are elliptical with a prominent, persistent style at the tip. There are 3-7 seeds in two rows in each fruit chamber.

Ecology & Ethnobotany This is an introduced species from Europe usually seen in old field and along roadsides. The genus name, *Berteroa*, is for Carlo Guiseppe Bertero, a Piedmontese botanist. The species name, *incana*, means gray. There is some information showing the importance of this plant species to wildlife. The plant produces many seeds, providing some feed value to songbirds and small mammals. Horses become intoxicated after eating green or dried plants. When mixed with alfalfa hay, *B. incana* can remain toxic for up to nine months. The toxic dose has not been determined.

BLACK MUSTARD

Brassica nigra & B. rapa   BRASSICACEAE

General Description *Brassica's* are large annuals with showy yellow flowers. The pods are round or 4-sided in cross section, with a conspicuous beak. In the Boise Foothills the species may be encountered in waste places and fields.

Comparison of *Brassica* species

B. **nigra** Stem leaves petioled or sessile, but not auriculate or clasping

B. **rapa** Stem leaves sessile and strongly auriculate-clasping

Ecology & Ethnobotany The genus is native to Western Europe, the Mediterranean and temperate regions of Asia. In addition to the cultivated species, which are grown worldwide, many of the wild species grow as weeds, especially in North America, South America, and Australia. Almost all parts of some species or other have been developed for food, including the root (swedes, turnips), stems (kohlrabi), leaves (cabbage, brussels sprouts), flowers (cauliflower, broccoli), and seeds (many, including mustard seed, oilseed rape). Some forms with white or purple foliage or flowerheads, are also sometimes grown for ornament.

LITTLE FLASE FLAX
Camelina microcarpa   **BRASSICACEAE**

General Description This is an erect, annual herb growing up to 28 inches tall. The lance-shaped leaves are mostly sessile and 3 inches long with entire or toothed margins. The leaves also have small lobes at the base that often clasp the stem. The flowers are pale yellow in color and the fruits are broadly elliptical in outline and rounded on top with an evident beak. This species occurs on sandy soils within the Boise Foothills.

Ecology & Ethnobotany A related species has been cultivated since the Neolithic Age for the fibers of its stem and the edible oil contained in the seeds. It is said that an oil similar to linseed oil.

SHEPHERD'S PURSE
Capsella bursa-pastoris   **BRASSICACEAE**

General Description This is an erect annual growing up to 20 inches tall. It has basal leaves that are dissected. The upper stem leaves become reduced in size and are more entire, lanceolate in shape and sessile with clasping base. The flowers are small, white in color. The fruits are heart-shaped or purse-shaped. This plant may be found on dry soils and disturbed areas in the Boise Foothills.

Ecology & Ethnobotany Shepherd's Purse is an easy plant to identify among members of the Mustard family because of the distinctive shape of its seedpods. This shape apparently resembles the leather purse of shepherds during the Middle Ages.

The nectar and pollen of the flowers attract mostly short-tongued bees and flies, including honeybees, Halictid bees, Andrenid bees, Syrphid flies, Tachinid flies, and flesh flies (Sarcophagidae).

Shepherd's purse has been used as food for thousands of years. The young leaves are edible raw or cooked. They are delicious when blanched. The plant is extremely high in Vitamin K, the blood clotting vitamin. Seeds can be ground into meal, or they can be used for flavoring other foods. Roots have been candied by boiling in syrup. Fresh or dried, they can be used in place of ginger. The pods are also useful as food.

TOOTHWORT, BITTERCRESS
Cardamine oligosperma & C. pensylvanica
BRASSICACEAE

General Description This large genus contains more than 150 species of annuals and perennials. The genus grows worldwide in diverse habitats. The flowers are white or purple and the pods are elongate and flattened. Two species may be found in the Boise Foothills. Several species were previously included in the genus *Dentaria*.

Ecology & Ethnobotany Some plants were reputed to have medicinal qualities (treatment of heart or stomach ailments). The name *Cardamine* is derived from the Greek word *kardamon*, referring to a Persian or Indian herb with pungent leaves.

WHITETOP

Cardaria draba & C. pubescens   **BRASSICACEAE**

General Description These two species are rhizomatous perennial herbs with erect or ascending stems and simple alternate leaves. Herbage is pubescent with simple hairs. The stalked flowers are borne on the upper portion of the stems in a crowded, more-or-less flat-topped inflorescence. The fruits are egg-shaped to nearly round with a persistent style at the top.

Ecology & Ethnobotany These species are aggressive perennials native to southwest Asia. They were likely introduced in multiple shipments of contaminated alfalfa seed from Turkestan into North America over a period of 40-50 years. All species readily establish in disturbed areas in range and wildlands and are favored during years of above average precipitation. Once established, colonies are difficult to eliminate because of deep, persistent roots. Cultivation can facilitate spread of plants by dispersing root fragments.

CROSSFLOWER

Chorispora tenella   **BRASSICACEAE**

General Description This annual species has stems growing up to 16 inches tall. The elliptical or lance-shaped leaves are 1 to 3 inches long, and mostly petiolate with coarsely toothed margins. Herbage is covered with simple and gland-tipped hairs. Flowers are borne on short stalks on the upper portion of the stems. The flowers are purple and the narrow and the fruit is curved. This weed is found along roads and trails, and in disturbed agricultural areas.

Ecology & Ethnobotany This plant is a native of Russia or southwest Asia and it was first documented in this country in Lewiston, Idaho in 1929. Since then, it has spread throughout the western plains states, the western portion of the United States, and southern Canada. Crossflower likely was introduced into the United States by accident in seed that was imported; as is true with many members of the mustard family.

TANSYMUSTARD
Descurainia   BRASSICACEAE

General Description Tansymustards are annual or biennial herbs with leaves that are 1-3 times pinnately divided. The foliage is covered with simple, branched, or short gland-tipped hairs. The flowers are cream-colored or light yellow and the pods are long, narrow, 3-sided to nearly round in cross section. The three species in the Boise Foothills are weedy and occur in disturbed soils.

Ecology & Ethnobotany During the 1950's uranium boom, botanists recognized that species of *Descurainia* were indicative of uranium and vanadium, which they absorb by growing in soils rich in these minerals.

The young green parts of these plants can be steamed for about half an hour; they are then eaten or dried for future use. Parboiling helps to remove the bitter taste. The seeds were parched by tossing in a basket with hot stones or live coals, then ground into a fine flour and made into mush. Because of its peppery taste, the mush was often mixed with the flour of other seeds to make it more palatable.

SPRING DRABA
Draba verna   BRASSICACEAE

General Description This is a delicate annual growing up to about 2 inches tall, sometimes up to 8 inches tall. The basal leaves are clustered, lance-shaped or narrowly spoon-shaped with entire or shallowly toothed margins. The plants are covered with branched hairs. The flowers are with white petals that are deeply lobed. The egg-shaped fruits are glabrous

Ecology & Ethnobotany The species is widespread and locally abundant in open soil. It is one of the first plants to bloom in the spring. They flower and yield mature fruits very rapidly and by the first of June have almost completely disappeared.

The genus is large and identifying species is difficult. Close inspection of hairs of the leaves is important in distinguishing the species, and worth the attention for the intricate forms hairs take.

WALLFLOWER
Erysimum   BRASSICACEAE

General Description Wallflowers are annual, biennial, or perennial herbs that are often tap-rooted. The herbage is covered with closely appressed forked hairs and yellow or orange flowers are showy. The linear pods are 4-sided in cross section with a small beak.

Sanddune wallflower (*E. capitatum*) is commonly encountered in the Boise Foothills.

Ecology & Ethnobotany As with most mustards, they are edible, but not choice. Wallflowers were once used as a poultice. *Erysio* means to draw out, as in drawing out pain or causing blisters. One of our species, *E. capitatum*, is a wide-spread, common, and highly variable species and commonly encountered in the Boise Foothills upwards into the forest.

PEPPERGRASS, PEPPERWEED
Lepidium BRASSICACEAE

General Description There are several species in the Boise Foothills including the native threatened slickspot peppergrass (*L. papilliferum*). The other species species non-native annual or biennial herbs with simple or 1-3 pinnately divided leaves that are alternate or basal. The flowers are white, yellow, or greenish, and the pods are flattened at right angles to the partition that separates the seed chambers. Look for these peppergrass in dry, open or vernally moist areas.

Ecology & Ethnobotany The young stems and leaves may be eaten raw or dried for future use. The plants contain Vitamins A and C, iron, and protein. The seed pods and seeds can be used as a flavoring.

Slickspot peppergrass is found only in small parts of the Boise Foothills. As its name suggests, this flower grows only where puddles or small pools form after rains or snow, and then dry up in the hot arid climate. In the foothills proper, there are a few small patches where these plants may be found (e.g., near Eagle and a very small patch near Columbia Village the author discovered in 2017). In general, the

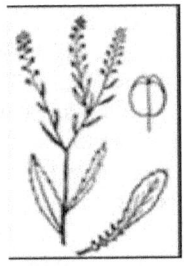

populations of this threatened plant have been reduced to a fraction of its former range. The primary threats being livestock trampling and grazing, off-road vehicles, agriculture developments, and other human activities. When found, inform the local land management agency.

WATERCRESS

Nasturtium officinale   BRASSICACEAE

General Description This is an aquatic perennial that is slightly juicy or succulent. Leaves are pinnate compound into 3 to 11 ovate leaflets. The flowers are white or yellow, and the fruit is a curved pod. Watercress is common in quiet streams or on wet banks below 8,000 feet. The small white flowers, succulent foliage, and aquatic habitat will immediately identify watercress. Flowers from March to November.

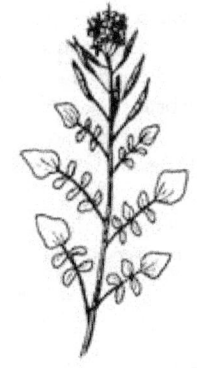

Ecology & Ethnobotany The peppery-tasting plants were eaten raw or cooked as a potherb. A good source of vitamins, watercress is listed as efficient in preventing scurvy. Watercress contains significant amounts of iron, calcium and folic acid, in addition to vitamins A and C. In some regions watercress is regarded as a weed, in other regions as an aquatic vegetable or herb. Where watercress is grown in the presence of animal waste, it can be a haven for parasites such as the liver fluke.

As noted, the herbage of watercress is edible if the waters in which they grow are not polluted. However, finding unpolluted water may be difficult. One suggestion would be to soak the fresh greens in a disinfectant, or treat the water with water purification tablets, or a tablespoon of bleach in a quart of water. Then rinse the greens well in potable water to remove the chemicals.

FENDLER'S PENNYCRESS
Noccaea fendleri   BRASSICACEAE

General Description This plant will form loose mats with few to several, simple or sparingly branched stems up to 8 inches tall. The basal leaves have lance-shaped or narrowly elliptical blades with entire or weakly toothed margins. The stem leaves are sessile with bases that clasp the stem. The plant is for the most part glabrous and often has a thin waxy coating. The flowers are white and the fruits are elliptic, cuneate, or oblanceolate in shape and flattened contrary to partition.

Ecology & Ethnobotany The species was once included in the genus *Thlaspi*. The genus name, *Noccaea,* was given in 1802 by German botanist Conrad Moench (1744-1805) to honor the Italian clergyman and botanist, Domenico Nocca.

BOG YELLOWCRESS
Rorippa palustris   BRASSICACEAE

General Description This is an annual or biennial plant with erect, mostly branching stems. The petiolate, lower leaves are elliptical in outline and pinnately divided with a terminal lobe larger than the laterals. Upper leaves are smaller, sessile, lobed or toothed, and clasping at the base. The plant in general is glabrous or hairy. The flowers are yellow or white; the pods are elliptical to linear and 3 sided to slightly compressed.

Ecology & Ethnobotany The leaves of most species are edible raw or after cooking. The plants are rich in iron, copper, calcium, sulfur, and magnesium. They also contain substantial quantities of Vitamins A, B, B2, and C.

TUMBLEMUSTARD
Sisymbrium altissimum   BRASSICACEAE

General Description Tumblemustard has erect, freely branched stems up to 5 feet tall. The plant is hairy at the base and glabrous above. The petiolate leaves are pinnately divided with numerous toothed or entire lateral lobes that are narrowly lance-shaped (below) to linear with a small, pointed lobe at the base. The petals are yellow petals and the spreading fruits are slightly 4-angled. They are about as broad as the stout stalks that bear them.

Ecology & Ethnobotany This species is a conspicuous roadside weed in our area. It also occurs in disturbed soil of fields and is widespread in North America. Tumblemustard is a honeybee and butterfly

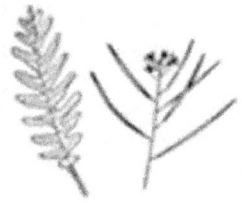

plant and it helps bind fine-textured soils. Native Americans made meal from ground tumble mustard seeds. The greens can be used in salads.

COMMON FRINGE-POD
Thysanocarpus curvipes   BRASSICACEAE

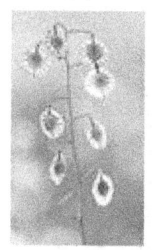

General Description This is a slender branched annual with stem leaves and basal leaves arranged in a rosette. The flowers are purplish, and the circular, flattened pods are surrounded by a flat nearly circular wing. The species occurs in open areas.

Ecology & Ethnobotany Fringe-pod seeds are edible after parching and being ground into flour. A tea made from the plant is said to cure stomach-ache; a drink made from the leaf can be used to relieve colic.

ROCKY MOUNTAIN BEEPLANT
Cleome serrulata   CAPPARACEAE

General Description This is an erect, showy plant up to 40 inches tall with alternate leaves divided into three lance-shaped, entire leaflets. The reddish-purple to pink flowers are arranged in a dense, narrow, terminal inflorescence. The petals are separate while the sepals are united. The fruits are long-stalked, pendulous capsules, linear to lance-shaped in outline. Beeplant is found in disturbed areas in the Boise Foothills.

Ecology & Ethnobotany An important food for many western Native Americans, bee plant was extensively used as a potherb. The young tender shoots and leaves, and flowers are preferred. The plant has an unpleasant odor, especially when older, and a pungent taste much like the mustards. We found it necessary to cook the plants in at least two changes of water to remove the bitter taste. The seeds can also be collected and ground into flour.

HONEYSUCKLE, TWINBERRY
Lonicera CAPRIFOLIACEAE

General Description The species here are shrubs or woody vines with entire and opposite leaves. In the Boise Foothills, Utah honeysuckle (*L. utahensis*) may be found in the upper elevations. The flowers are quite fragrant and we've usually smelled the flowers before actually seeing it.

Ecology & Ethnobotany The berries of this species are edible and can be eaten raw or dried for future use. The long stems of honeysuckle can be used as basket foundation material. You can also peel and split the hairy stems and use as wrapping material for coiled baskets.

ELDERBERRY
Sambucus nigra   CAPRIFOLIACEAE

General Description Elderberries are shrubs with pithy stems. This species has large, compound leaves with serrated leaflets, the white flowers are arranged in dense clusters, and the fruits are blue-black in color.

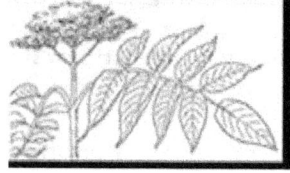

Elderberries can be found in open areas, hillsides, and riparian habitats in the Boise Foothills.

Ecology & Ethnobotany The blue or black colored berries are edible raw, or they can be made into excellent jams, jellies, and wine. They can also be dried and stored for future use. The seeds contain hydrocyanic acid, and if eaten in quantity can cause diarrhea and nausea. It is best to cook the berries or strain the seeds before use.

SNOWBERRY
Symphoricarpos albus & S. oreophilus
CAPRIFOLIACEAE

General Description Snowberries are erect shrubs with elliptical to egg-shaped leaves. Flowers are white to pink and bell-shaped, accompanied by two small bracts. The fruits are berry-like and white. Two species are found in the Boise Foothills in dry soils.

Comparison of *Symphoricarpos* species

S. oreohilus Corolla relatively long and narrow, an elongated bell-shape, longer than wide; stems with solid pith

S. albus Corolla short, broad and short-campanulate, scarcely, if ever, longer than wide

Ecology & Ethnobotany The white, tasteless berries are edible raw or cooked, and are said to be emetic and cathartic in large amounts. Saponins are found in the leaves and can be used as a natural cleaning agent. The new twigs are flexible and can be used in cordage and basketry.

BALLHEAD SANDWORT
Arenaria congesta   CARYOPHYLLACEAE

General Description Ballhead sanduort has slender, erect stems up to 12 inches tall from a branched, often woody caudex. The sharp-pointed leaves are glabrous or sparsely hairy on the margins and up to 3 inches long. There are 2-4 pairs of stem leaves, and only the basal leaves have secondary leaves in the axils. The flowers are borne in dense to open inflorescences. Sepals have a broad, membranous margin and are blunt to acute at the tip. The petals are longer than the calyx. This species common in dry, open habitats in the upper reaches of the Boise Foothills.

Ecology & Ethnobotany Thomas Nuttall named this species *Arenaria congesta* in 1838 and S. S. Ikonnikov renamed it *Eremogone congesta* in 1973. *Congesta* refers to the crowded flower head.

MOUSE EAR CHICKWEED
Cerastium fontanum   CARYOPHYLLACEAE

General Description This is a biennial or short-lived perennial with lax or ascending stems. It goes to about 16 inches tall and often roots at the nodes. The lower leaves are oblong to broadly lance-shaped and the upper leaves are larger. The herbage is densely glandular and hairy throughout. The sepals have whitish, papery margins. The petals are shorter than the sepals. This is a common plant of waste places at lower elevations.

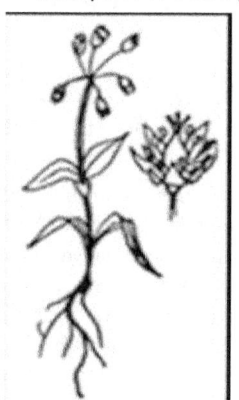

Ecology & Ethnobotany The species is introduced from Eurasia and is now widespread in North America. *Cerastium* is frequently confused with *Stellaria media* (chickweed), but to the general forager there is no danger. The tender leaves and stems of most *Cerastium* can be added to a salad, but we found they are better if boiled first and served as greens.

JAGGED CHICKWEED
Holosteum umbellatum   CARYOPHYLLACEAE

General Description This annual has simple or basally branched stems growing less than 8 inches tall; usually much smaller. The basal leaves are oblong lance-shaped and the 2-3 pairs of stem leaves are wider. The foliage is soft hairy below and glandular hairy above. Flowers are borne in a flat-topped inflorescence at the top of the stem. Sepals are separate and the white petals are jagged at the tips and slightly longer than the sepals. The seed capsule is nearly twice as long as the sepals.

Ecology & Ethnobotany Jagged chickweed is common in moist open habitats. In years with a wet spring, it appears in great masses in overgrazed pastures. Introduced from Eurasia and established throughout much of North America.

ROSE CAMPION
Lychnis coronaria   CARYOPHYLLACEAE

General Description This is a perennial with several unbranched stems arising from a branched rootcrown. The leaves are lance-shaped and they lack petioles and become smaller and spaced further apart upwards on the stem. The herbage is covered with dense, whitish hair. The calyx is enlarged in fruit, and the lobes are twisted. The petals are deep red with a heart-shaped blade.

Ecology & Ethnobotany This species is a garden flower native of Europe. It has escaped and persists in old fields and other disturbed areas. Species of *Lychnis* are similar to members of the genus *Silene* but can be distinguished by having 5 styles rather than 3.

LARGELEAF SANDWORT
Moehringia macrophylla   CARYOPHYLLACEAE

General Description This species has narrowly elliptic to lance-shaped leaves, usually less than 2 inches long and acute at the tip. The herbage is lightly hairy and rough to the touch. The 2-5 flowers are borne in an open inflorescence. Sepals are 1/8-inch long and pointed at the tip, and petals are usually about as long as the sepals. The capsule is oval or round. Largeleaf sandwort is common in moist or, more commonly, dry, often-forested habitats.

Ecology & Ethnobotany William Hooker named this plant *Arenaria machrophylla* in 1830 and then Eduard Fenzl renamed it to *Moehringia macrophylla* in 1833. The plant is named for Paul Moehring, an 18th century German physician, botanist, and zoologist.

TUBERED STARWORT
Pseudostellaria jamesiana   CARYOPHYLLACEAE

General Description This is a weak-stemmed, glandular perennial with lanceolate leaves and many few-flowered cymes of white flowers that have slightly 2-lobed petals. In the Boise Foothills, this species is found in damp places and flowers from May to July.

Ecology & Ethnobotany *Pseudostellaria* refers to starwort's resemblance to the genus *Stellaria*. The tuber-like swellings of this species can be eaten raw or dried in the sun. They have a thin, light brown rind, and a tender rather mealy texture inside, similar to a potato.

BOUNCING BET

Saponaria officinalis   CARYOPHYLLACEAE

General Description This an erect perennial herb with sessile or nearly sessile leaves. The flowers are showy, usually pale pink. Bouncing bet can be found along roadsides, disturbed areas, and waste places at the lower elevations. The plant has escaped from cultivation

Ecology & Ethnobotany The genus name is Latin for soap, since the juice of the plant lathers with water. The plant contains saponins and will irritate the digestive tract if eaten. The crushed green plant and roots can be used as a soap substitute.

MENZIES' CATCHFLY

Silene menziesii   CARYOPHYLLACEAE

General Description This is a low-growing, often matted plant with numerous lax or ascending stems arising from slender rootstocks. The leaves are lance- to egg-shaped, acute at the tip, and tapered to the base. Flowers are borne on slender stalks. The calyx is tubular to bell-shaped, obscurely nerved and the petals are white, and the blade is deeply cleft with 2 small appendages at the base of the blade.

Ecology & Ethnobotany The species is found in moist, partially shaded habitats. This plant is sometimes confused with *Moehringia macrophylla*. The two plants are best separated when in flower: *S. menziesii* has united versus separate sepals and its petals are narrow at their base.

PURPLE SAND SPURRY
Spergularia rubra   CARYOPHYLLACEAE

General Description This is a small, freely-branched annual with prostrate stems. The linear to fiiliform leaves are clustered at the papery, white-bracted nodes. The foliage is hairy and glandular. The flowers are borne in open, terminal, leafy-bracted inflorescences. The purplish-red petals are shorter than the sepals and open only in direct sunlight. This plant is common in dry, open disturbed habitats.

Ecology & Ethnobotany The tiny seeds are sometimes eaten but contain saponin. They were gathered, ground up, and mixed with flour to make bread.

CHICKWEED
Stellaria media   CARYOPHYLLACEAE

General Description Common chickweed is a slender, weak-stemmed annual with trailing stems. The leaves are opposite and the flowers are small and white. Looking closely at the petals, at first it appears as though there are 10 petals, but in actuality there are only 5; the

petals are what we call bifid. These is a rather common plant in moist shady areas along the trail, especially parks and lawns.

Ecology & Ethnobotany While the uses of other *Stellaria* are unknown, the young shoots of common chickweed have been used as salad herbs or potherbs if cooked like spinach. Although it is edible raw, we prefer to boil for a few minutes before eating. Since the plants are usually quite small and only the youngest parts are good, chickweed can be tedious to collect. The greens are low in calories and packed with copper, iron, phosphorus, calcium, potassium, and Vitamin C - valued in the prevention and treatment of scurvy.

FOURWING SALTBUSH
Atriplex canescens CHENOPODIACEAE

General Description *Atriplex* are annual or perennial herbs or shrubs with alternate leaves, and glabrous or scaly herbage. The flowers are unisexual, and individual plants have one or both sexes.

Ecology & Ethnobotany The many uses for these plants include food to medicine and dyes, as well as soap and spice. The young leaves of many species can be cooked and eaten as greens and have a very distinct salty taste. We've often added them to otherwise bland foods to make our wild meals less boring. Add the leaves to meats while cooking will help spice them up. The seeds were parched, ground into flour, and made into mush. They can also be soaked in water for a few minutes to make a rather pleasant tasting drink. The Navajo used the flowers to make puddings. The ashes of

A. canescens make a good substitute for baking soda.

The leaves and roots of many species were used as soap. They were rubbed in water for lather and used in washing clothing and baskets. Many Native Americans also carved arrowheads from the wood for use as weapons and for hunting. The seeds of some species were also used in making a black dye.

COMMON KOCHIA
Bassia scoparia   CHENOPODIACEAE

General Description Formerly in the genus *Kochia*. Common kochia is a bushy annual with stems up to three feet tall. The leaves are alternate, narrowly lance-shaped and tapered at both ends. The herbage may or may not be covered with hairs. Flowers are solitary or in clusters in spikes. The species is common in open, disturbed habitats at low elevations.

Ecology & Ethnobotany Common kochia is native to Europe and was introduced into the United States as an ornamental. It has since escaped and has become well established. In Asia, Japan, and China, common kochia was cultivated for its seeds. The tips of the young shoots can be prepared as potherbs. The seeds can be eaten raw or cooked, or ground into meal and used in bread making.

GOOSEFOOT
Chenopodium   CHENOPODIACEAE

General Description The species in this genus are annuals with mealy or glandular foliage. Leaves are alternate and entire, toothed, or lobed. Flowers are borne in dense clusters in the leaf axils or terminal inflorescence. There are 2 to 5 green or red sepals. Three species in the area include lambsquarters (*C. album*), blite goosefoot (*C. capitatum*), and Fremont's goosefoot (*C. fremontii*).

Ecology & Ethnobotany Leaves, tops, and seeds of all species can be used as an emergency or basic food and are quite tasty and nutritious. High in protein, the greens are a good source of Vitamins A and C, iron, potassium, and are extremely rich in calcium. Since it does not become bitter  with age, both young and old plants can be used. Leaves may be used raw in salads or boiled in water like spinach. The water can be saved and used as a yellow dye. The leaves were also eaten to treat stomachaches and prevent scurvy. A leaf poultice was used on burns. The flower buds and flowers can be used as potherbs. A single plant can produce up to 70,000 seeds. Seeds can be ground as flour for use in bread or cooked as mush. Seeds can also be eaten without grinding or incorporated into pinole (flour made from a mixture of seeds of small plants). The seeds contain about 15% protein and 55% carbohydrates, more than is found in corn. The seeds can also be used as a coffee substitute.

JERUSALEM OAK GOOSEFOOT
Dysphania botrys   CHENOPODIACEAE

General Description Formerly known as *Chenopodium botrys* and as *Ambrosia mexicana*. The genus *Dysphania* is known as the glandular goosefoots. Jerusalem oak goosefoot is native to the Mediterranean region.

Ecology & Ethnobotany The leaves and seeds are more or less edible. Leaves can be cooked or raw leaves should only be eaten in small quantities. The seeds can be ground into a meal and used with flour in making bread. The seed is small and fiddly, it should be soaked in water overnight and thoroughly rinsed before it is used in order to remove any saponins. The leaves are a tea substitute.

A gold or green dye can be obtained from the whole plant. Additionally, the dried plant is a moth repellent. The whole plant is very aromatic and is used as a scent in pillows, bags, and baskets.

SPINY HOPSAGE
Grayia spinosa   CHENOPODIACEAE

General Description This is a low mealy appearing shrub with stiff, spreading spine-tipped branches and slightly fleshy leaves with gray tips becoming pinkish with age. The flowers occur in heads, these borne in terminal or axillary spikes or panicles. The fruits are closely subtended by a pair of attractive rose-purple, thin, flat-winged bracts that are united to the middle or higher.

Ecology & Ethnobotany Some Native Americans ground parched seeds of spiny hopsage to make pinole.

WINTERFAT

Krascheninnikovia lanata   CHENOPODIACEAE

General Description Winterfat is a small shrub found at the lower plains and foothills elevations, often in saline or alkaline areas. The leaves are alternate, narrow and entire, whereas the flowers occur in heads or spikes in the axils of the leaves.

Ecology & Ethnobotany Winterfat is an important forage plant for horses and other livestock. Medicinally, the plant has been used by many Native American tribes. For example, the Hopi Indians used the powdered root for burns, and a decoction of the leaves was used for fevers. The Navajo made a poultice of the chewed leaves and applied it to a poison ivy rash.

NUTTALL'S POVERTYWEED

Monolepis nuttalliana   CHENOPODIACEAE

General Description Povertyweed is a low growing winter annual with prostrate or ascending stems. The leaves are somewhat succulent and lance-shaped, broadened and lobed at the base. Flowers are borne in dense clusters at the leaf bases and the solitary sepal is reddish in color. The seeds are dark brown. The plant can be found in open disturbed habitats at the lower elevations.

Ecology & Ethnobotany The above ground parts of povertyweed may be eaten as a potherb. The seeds are also edible.

PRICKLY RUSSIAN THISTLE
Salsola tragus   CHENOPODIACEAE

General Description This is not a true thistle (*Cirsium*), but a many branched annual with purplish striped stems up to three feet tall in a rounded form. The lower leaves are threadlike; the upper leaves are awl-like and spine-tipped. The plant may or may not be hairy. Flowers are solitary in the leaf axils and are subtended by spiny bracts.

Ecology & Ethnobotany Russian thistle is common in open, disturbed habitats, particularly around agricultural areas at low elevations. It was introduced to the United States from Europe. It is considered a noxious weed because of its distributional pattern and spines.

When mature, the whole plant becomes rigid, breaks off at ground level, and becomes one of the "tumbleweeds" that blows across the open plain.

This unsavory looking plant is edible. The young parts of the plant may be boiled and eaten as a potherb or chopped raw into a salad. On older plants, clip the tender branch tips that are green. In Europe, the ashes of the plant were once used in the production of carbonate of soda known as Barilla. **Warning** The older parts of the plants contain significant quantities of nitrates and oxalates and may be toxic if eaten in quantity.

GREASEWOOD

Sarcobatus vermiculatus **CHENOPODIACEAE**

General Description This perennial native is a long-lived shrub with spreading, rigid branches that often bear spines. The leaves are linear, succulent, and pale green with entire margins. Some of the leaves may be opposite and some alternate, but all the leaves are shed in winter. The plants usually have both male and female flowers on the same plant (monoecious), but they can occur on separate plants (dioecious). Male flowers are catkin-like spikes on the ends of branches and the female flowers usually occur singly in the axial of leaves and form the fruits that are surrounded by a green membranous wing.

Ecology & Ethnobotany Greasewood is used as wood for fuel and the sharpened spines were used for painting by Native Americans. Native Americans used the seeds and leaves, which have a salty taste, for food.

Seeds, leaves, and new leaders are also consumed by a variety of small mammals. It is an important browse plant and is rated from good to useless forage for cattle, sheep, and big game animals in the winter and provides good cover and food for small mammals and birds.

ST. JOHN'S WORT

Hypericum perforatum & H. scouleri **CLUSIACEAE**

General Description These species have yellow flowers, and small, translucent glands on the leaves and petals. The species can be found in moist areas at various elevations. Two species may be encountered in the Boise

Foothills: the creeping St. John's wort (*H. anagalloides*) and the erect species Scouler's St. John's wort (*H. scouleri*).

Ecology & Ethnobotany In general, Native Americans dried the whole plants and pulverized them into a meal, which was then used in cooking. The fresh leaves were also eaten. **Caution** The genus contains at least 6 species, worldwide which are poisonous. The toxin is a pigment, hypericin, which causes photosensitivity in the skin of animals who eat it. Upon exposure to light, such skin will form lesions, seep, itch, or fall off in more severe cases. There is no complete recovery from the after -effects of this poisoning.

FIELD BINDWEED
Convolvulus arvensis   CONVOLVULACEAE

General Description This is a perennial with trailing or twining stems. It spreads from a deep and brittle rhizome. The flowers are white or pinkish, funnel-shaped arising from the axils of the arrowhead-shaped leaves.

Ecology & Ethnobotany This beautiful, but

pernicious European weed is well established throughout North America and is frequently encountered on roadcuts and fields at low elevations. It is difficult to eradicate because of its deep rhizome and low growth.

RED-OSIER DOGWOOD
Cornus sericea   CORNACEAE

General Description Dogwoods are shrubs or semi-woody perennials with simple leaves that are opposite or whorled. A distinguishing characteristic of this family is the flower structure, which includes 4 to 5 sepals, petals, and stamens, all of which are attached at the top of the ovary. The flowers mature into red or white drupes.

Red-osier dogwood is a shrub that occurs in many moist habitats and is highly variable with many local forms. The berries, which resist rot due to their low sugar content, provide long-lasting food for wildlife during winter.

Ecology & Ethnobotany The most distinctive feature of this plant is its bright red bark. The Native Americans scraped, dried and smoked the inner lining of this bark, which is reported to have hallucinatory properties. The term osier refers to pliable twigs used in basket making.

The wood of this family is extremely hard and free of scratchy silica - so much that jewelers reportedly used small splinters of it to clean out the pivot-holes in watches and opticians utilized it to remove dust from small-deep-seated lenses. Peeled twigs of any species can be used as tooth brushes.

STONECROP
Sedum lanceolatum & S. stenopetalum
CRASSULACEAE

General Description Stonecrops are well-adapted to survival in shallow soil or on rocky outcroppings. The succulent leaves and stems have a waxy coating to help reduce water loss. The reddish color of the foliage in some species is enhanced by sunlight and occurs most often in plants in hot exposed sites. Two species of *Sedum* can be found in the Boise Foothills.

Ecology & Ethnobotany These are my favorite edible plants. The young leaves and stems of all species can be eaten as a salad or boiled as a potherb. I find them slightly tart and crisp - a wonderful addition to salads and trail snack. However, some species have emetic and cathartic properties, and can cause headaches. In an emergency, stonecrop can be eaten raw to allay hunger and thirst. The plants are best when collected before flowering since they tend to become bitter and fibrous in late summer. The green fleshy leaves are high in Vitamins A and C. The tubers can also be boiled and eaten. **Caution** *Sedum* has also been reported as being slightly astringent and mucilaginous.

DODDER

General Description Members of the dodder family are leafless, rootless, parasitic herbs that lack chlorophyll. The stems are thread-like and often yellowish in color. The small flowers have 4 or 5 distinct sepals, and 4 or 5 united petals. The fruit is a dry or fleshy globose capsule. The genus is native to the U.S. and may cause great losses to crop plants. The family was once included in the Convolvulaceae (Morning Glory Family).

Ecology & Ethnobotany Dodders are leafless, twining perennials with slender stems that colored pink, whitish, or yellowish, never green. Both the leaves and pink to white flowers are highly reduced. Dodder can normally be identified only with a microscope or hand lens. The many species of *Cuscuta* parasitize different flowering plant hosts at low elevations.

Dodders have a very unique life cycle. The small seeds usually germinate in the soil and produce slender stems without seed leaves (cotyledons). Unless the slowly rotating plant encounters a host plant within a short period of time, the dodder seedling will wither and die. However, if the seedling encounters the living stem of a susceptible host plant, the dodder will twine around it and at certain points develop suckers that penetrate the tissue of the host. Nutrition is received through these suckers. Dodder then losses all contact with the soil. After a period of growth, small flowers develop and large amounts of seeds are produced to start the process all over again.

SEDGE
Carex   CYPERACEAE

General Description Sedges are perennial, grass-like plants with creeping rhizomes, short rootstalks, or fibrous roots. The solid (not hollow) stems are usually triangular in cross section, and the leaves at the base of the stem are often reduced to scales. Carex is one of the largest genera in this area. Many of the species are difficult to tell apart, and a more technical key should be consulted. Mature fruit and a hand lens are necessary for positive identification of most species.

Ecology & Ethnobotany Any sedge should be tried for food. The young shoots or bases of the leaves are tasty additions to the diet

FULLER'S TEASEL
Dipsacus fullonum   DISPSACACEAE

General Description Fuller's Teasel is a stout, prickly tap-rooted plant up to six feet tall. The leaves are opposite and lance-shaped and distinctly prickly on the lower surface of the midrib. The small, bluish-purple

flowers occur in a large, terminal flower head that is egg-shaped and armed with numerous, sharp-pointed bracts.

Ecology & Ethnobotany This non-native plant from Europe is found throughout North America on disturbed soils with appreciable water holding capacity. The plant has a tendency to be opportunistic and displaces desirable native plants.

Cloth cleaners use the dried flower heads to remove the nap from wool or cloth after beating and cleaning. Apparently, the teasel heads perform the task so well that man-made tools have not replaced them.

BRITTLE BLADDERFERN
Cystopteris fragilis   DRYOPTERIDACEAE

General Description This 6-16 inches tall fern is loosely tufted from a short creeping rhizome. The leaves are thin and delicate in texture. The stipes are brown below, yellowish above, and smooth. The indusia are small, attached at one side and arching back to form a hood. This is a widely distributed fern and is found in the crevices of cliffs and ledges, in soil under rocks, shrubs, or trees.

Ecology & Ethnobotany The genus has been placed in the family Cliff Fern family – Woodsiaceae. The plant was used as a dermatological aid by the Navajo. Here, a cold, compound infusion of the plant was made and used as a lotion for injuries.

OREGON CLIFF FERN
Woodsia oregana   DRYOPTERIDACEAE

General Description This small fern commonly grows in rocky places. The underground stem is densely tufted and clothed with broad, thin scales. The leaves are clustered, numerous, small, linear to lanceolate-ovate, and once- or twice-pinnate. The sori are round and seated on the back of the free veins, and the indusia is under the sori with star-shaped divisions.

Ecology & Ethnobotany The genus name honors Joseph Woods, an English botanist. Rocky Mountain woodsia was used as a sign of water when traveling through the mountains by the Natives further to the north. Some species are cultivated as ornamentals. Some authorities have placed the genus in the family Cliff Fern family – Woodsiaceae.

HORSETAIL
Equisetum   EQUISETACEAE

General Description In general, these are rhizomatous ferns with hollow, grooved, regularly jointed stems that are impregnated with silica. The leaves are reduced in size, appearing as a series of teeth around a joint. Spores are produced in cone-like structures atop the stems. They are found in moist soil along streams and rivers, marshes, and other damp habitats. At least three species may be encountered in the Boise Foothills.

Ecology & Ethnobotany Although all species are useful and identical in application, common horsetail (*E.*

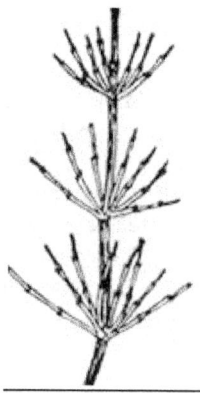

arvense) is the most popular. The tough outer tissue can be peeled away and sweet inner pulp of all species can be eaten in small amounts. In large quantities, defined as greater than 20% of body weight by some authorities, they can be toxic. Certain chemicals in this plant are said to destroy specific B Vitamins such as thiamine. The enzyme thiaminase is apparently responsible for the poisoning.

Cooking destroys this enzyme and renders the plants safe for consumption. The tuberous growth on the roots (actually rhizomes) can be eaten raw in the early spring or boiled later in the season.

In the fall, the stems become impregnated with silicon dioxide and can be used to scour pots and pans or as a type of sandpaper for wood.

SPURGE & SANDMAT
Chamaesyce & Euphorbia   EUPHORBIACEAE

General Description These are annual or perennial herbs with milky juice. The flowers are borne in a complex structure called a cyathium. This cup-like structure contains several male flowers and a single female flower. They are found in disturbed habitats in low elevations.

Ecology & Ethnobotany The two genera are similar and have the cyathia-type flower arrangement. In fact, all the species discussed here were classified as *Euphorbia*. In short, species of *Euphorbia* usually have alternate leaves and copious milky sap. In contrast, *Chamaesyce* have opposite leaves and white sap.

Euphorbia contains toxic principles that will cause severe poisoning if ingested in quantity. Most species contain carcinogenic, highly irritant, diterpene esters and are strong purgatives. The white sap can cause skin irritations and blisters.

TURKEY-MULLEIN
Croton setiger   EUPHORBIACEAE

General Description Formerly known as *Eremocarpus setigerus*. turkey-mullein is a low spreading gray plant with hispid stems and pubescent leaves. Leaves are thick, ovate in shape and palmately veined. This is an abundant weed encountered in disturbed areas in the Boise Foothills.

Ecology & Ethnobotany The herbage of the plant is regarded as poisonous. The poison, presumably diterpenes, was used to stupefy fish in the past.

MILKVETCH
Astragalus   FABACEAE

General Description These are perennial herbs mostly with leaves that are pinnately divided into several or numerous leaflets. There are small, membranous or leafy appendages at the juncture of the leaf with the stem (stipules). The foliage is glabrous or covered with simple hairs that are attached at the base (basifixed) or near the middle (dolabriform) and thus appearing 2-branched. The few or numerous, pea-like flowers are borne on short stalks of narrow inflorescences that arise from the axils of the upper leaves. The upper petal (banner) is usually the longest, and the united lower petals (keel) are rounded or broadly pointed. There are 10 stamens. Pods are variously shaped and textured and are often at least partially divided into two seed chambers.

Ecology & Ethnobotany This is a difficult genus of perhaps 1600 species, making it the largest genus in the Pea family.

Mulford's milkvetch (*A. mulfordiae*) is a sensitive species that may be encountered in the Boise Foothills. The conservation threats include habitat destruction due to urbanization, and habitat degradation, especially weed invasion, from wildfires, livestock grazing, mining, off-highway-motorized vehicles, and nonmotorized recreational activities. The combination of a limited distribution, usually small population size, and vulnerability to ongoing threats means that the species is a high priority conservation concern in the area.

LICORICE
Glycyrrhiza lepidota   FABACEAE

General Description This plant has deep, extensive rhizomes which give rise to erect, rigid stems growing up to 48 inches tall and that are covered with stalked or sessile glands. Yellow-brown glands also dot the underside of the 11-19 lance-shaped leaflets. The pale yellow to greenish-white flowers are crowded along the terminal potion of the long, naked stalk arising from the upper leaf axils. The short pods are brownish and have dense bristles. This species is usually found in moist, sandy soils.

Ecology & Ethnobotany The plant contains glycyrrhizin, sugar and other chemicals used in medicine as a mild laxative, a demulcent, and a flavoring to mask the taste of other drugs. It is also used in confections, root beer, and chewing tobacco. Licorice has been used in the treatment of asthma, stomach ulcers, bronchitis, and urinary tract disorders. The plants were chewed by Native Americans and used as a flavoring.

LUPINE
Lupinus   FABACEAE

General Description There are few species species of lupine in the Boise Foothills. In short, they are showy perennial or annual herbs with palmately compound leaves. Flowers range in color from blue, violet, rarely white, to rose and occur in elongated narrow inflorescences. The pods are flattened and usually hairy. They are found on open slopes and meadows up into the alpine zone.

Ecology & Ethnobotany Linnaeus named this genus in 1753. *Lupinus*, which is Latin for Wolf, was so named because of the erroneous belief that the species degraded land.

Due to their abundance in some areas, it is **important to quickly dismiss these plants as being edible.** In many field guides, the pea-like seeds have been wrongly recommended by the authors as a substitute for peas. Lupines possess many complex alkaloids and should be considered poisonous.

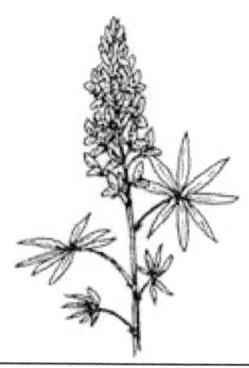

BLACK MEDICK, ALFALFA
Medicago lupulina & B. sativa   FABACEAE

General Description In general, they are described as hairless, branching perennial or annual herbs with leaves divided into three leaflets. The terminal leaflet evidently longer than the other two. The pods are twisted. Usually found in disturbed areas at the lower elevations in the Boise Foothills.

Comparison of Medicago species

M. lupilina Flowers less than $\frac{1}{4}$ inch long; shallow rooted annual
M. sativa Flowers $\frac{1}{4}$-$\frac{1}{2}$ inch long; deep-rooted perennial

Ecology & Ethnobotany The dried and powdered young leaves and flower heads of alfalfa are nutritious and can be steeped in hot water to make a bland tea. Alfalfa sprouts are a popular salad addition and the seeds are available from various health stores. Nectar from the flowers produces a good honey.

In addition to uses as food and medicine, alfalfa seeds contain an oil for use in paints and varnishes. Papermakers have used the stem fibers in their craft, and wool dyers extract a yellow dye from the seeds.

YELLOW SWEETCLOVER
Melilotus officinalis   FABACEAE

General Description Sweetclovers are strongly tap-rooted perennial or annual herbs. The leaves are divided into three fine-toothed wedge-shaped leaflets. The white or yellow flowers are loosely arranged in an inflorescence and the pods are thickly spindle-shaped. The plants are usually found in disturbed habitats within the Boise Foothills.

Ecology & Ethnobotany The young leaves (before the flowers appear) may be eaten raw or boiled. The fruit may be used as seasoning for soups. The older leaves are toxic and should be avoided. Improperly dried sweetclover will easily mold and should be considered poisonous. Molding sweetclover mixed in hay has killed many cattle.

CLOVER
Trifolium   FABACEAE

General Description There are many species of clover in the Boise Foothills. In general, they are annual and perennial plants from rhizomes with leaves that are divided into three or more leaflets. The flower colors range from white, pink, yellow, red, or purple, and the seed pods are round to elongated. They are found in various habitats. Two common non-native species found in disturbed habitats include red clover (*T. pratense*) and white clover (*T. repens*).

Ecology & Ethnobotany
All species are nutritious and high in protein, but the flower heads and tender young leaves are hard to digest raw and may cause bloating. To improve digestibility of the plants, soak them in salt water for several hours or overnight. Leaves prepared this way may be dried and stored for future use. The dried flower heads and seeds can be ground into a flour substitute or extender.

VETCH
Vicia   FABACEAE

General Description Vetch are annual or perennial herbs with trailing to climbing stems. Leaves are pinnately divided with tendrils in place of terminal leaflets. Three introduced species can be found growing in waste places within the Boise Foothills. The native species (*V. americana*) is also found along roadsides, fields, and open places.

Ecology & Ethnobotany *Vicia* closely resembles *Lathyrus* and requires close examination of the stipules. The stipules of *Vicia* are usually cut into narrow lobes, whereas the stipules of *Lathyrus* are entire to dentate. *Both genera should be avoided and considered poisonous despite some people suggesting otherwise.*

LONGHORN STEER'S-HEAD
Dicentra uniflora   FUMARIACEAE

General Description Plant is very small and ground-hugging. A single flower (may be pale pink, lilac, or white in color) is borne at the tip of each leafless stem. The flowers are small, about ½ inch long. The corolla with a cordate base and outer petals that are slightly pouched.

Ecology & Ethnobotany Because of its diminutive size and early blooming, steer's-head appears to be less common than it is. It is found on well-drained sites in the Boise Foothills.

The plants are considered to be poisonous and contain several different alkaloids. These alkaloids are found throughout the plant and can cause trembling, staggering, convulsions, and labored breathing. Large quantities can be fatal.

WHITESTEM FRASERA, WHITE FRASERA
Frasera albicaulis & F. montana   GENTIANACEAE

General Description These are medium-sized plants with large, terminal clusters or spikes of flowers. The corolla lobes have distinct fringed glands at their bases. Some species have a circle of hair-like scales surrounding the stamens.

Comparison of *Frasera* species

F. albicaulis Corolla lobes bluish
F. montana Corolla lobes creamy white

Ecology & Ethnobotany The genus is named for John Fraser, a Scottish botanist. Historically, *Frasera* has sometimes been considered part of the genus *Swertia*, but molecular analysis has shown them to be separate from *Swertia*.

The genus *Frasera* differs from *Swertia* in having a slender style and the flower parts occur in fours. *Swerta* has a relatively thicker style and the flower parts occur in fives.

RED-STEMMED STORKS-BILL
Erodium cicutarium   GERANIACEAE

General Description Red-stemmed storks-bill is a low growing annual with mostly basal, finely dissected, fernlike, pinnately divided leaves. The flowers are small, pink and mature into the distinctive "stork's bill" fruit. This is an introduced plant that is widespread on disturbed sites within the Boise Foothills.

Ecology & Ethnobotany The leaves of this species can be eaten raw in salads or cooked as a potherb. They are particularly palatable when picked young and have a parsley-like taste. We find it nicely compliments an otherwise bland wild salad and provides a good source of Vitamin K. It is uncertain whether other species of *Erodium* are edible and it is not recommended.

STICKY PURPLE GERANIUM
Geranium viscosissimum   GERANIACEAE

General Description Sticky geraniun is stout, hairy, and grows up to 3 feet tall. Large basal leaves are up to 12 inches long. There are glandular hairs on the sepals and flower stalkswhich are yellow-tipped. Flower

color is variable, mostly purplish-red to pinkish-lavender. The inner petal surfaces have long, straight hairs. Sticky geranium is a a common plant of lower elevation grasslands, generally associated with lupines and balsamroot. It also occurs in wet meadows and open, drier forest types, but it can be found all the way to timberline.

Ecology & Ethnobotany The leaves and flowers of most *Geranium* species can be eaten, but because of their astringent properties and texture, they are not a choice edible. We find that they are best when tossed in with other greens in salads or steamed as potherbs. In any case, the leaves are better treated as a filler to stretch supplies of tastier and less abundant greens. The leaves can also be chopped and added to soups, thereby blending flavors making the leaves more acceptable.

Geranium leaves are similar looking to monkshood (*Aconitum*), so positive identification of the flowerless plants is important.

CURRANT, GOOSEBERRY
Ribes   GROSSULARIACEAE

General Description Members of this genus are shrubs. The species that have prickles on the stems and bristles on the fruit are commonly called gooseberries. Those without prickles on the stem or bristles on the fruit are currants. Leaves are palmately veined and shallowly or deeply lobed. The five petals are smaller than the sepals and usually narrowed to a claw-like base. Fruit is a berry.

Ecology & Ethnobotany The berries of all species are edible raw, and none are known to be poisonous. However, we have come across some unpalatable species, berries with an unpleasant odor and a taste to match. The berries are high in Vitamin C and one of the richest plant sources for copper. One method of collecting them in bulk, is by shaking the bushes over sheets of plastic or blankets. Those that are too sour or spiny become more palatable if they cooked or dried. In regard to the fruits with bristles, one can also roll the berries on hot coals in a basket until the bristles have been

singed off. When dried, the berries are a great trail snack. The dried berries can also be mixed with meat to make pemmican. The berries contain enough natural pectin to make jelly. The seeds also contain large quantities of gamma-linolenic acid and many herbalists use this oil to treat skin conditions, asthma, arthritis, and premenstrual syndrome. The nectar-filled flowers are considered good trail snacks. The wood makes good arrow shafts.

LEWIS' MOCKORANGE
Philadelphus lewisii   HYDRANGEACEAE

General Description This is the State Flower of Idaho and is also known by the common name of Syringa. This is a loosely branched shrub or tree with white flowers. It grows on rocky slopes and canyons below 6,000 feet. The genus is named for the Egyptian King Ptolemy Philadelphus.
Ecology & Ethnobotany The wood of mockorange is strong and hard and does not crack or warp. It is an excellent wood for making bows and arrows. The leaves and flowers foam into lather when bruised and rubbed with hands and can be used for cleaning the skin. The plant is otherwise considered poisonous.

BALLHEAD WATERLEAF
Hydrophyllum capitatum   HYDROPHYLLACEAE

General Description Ballhead waterleaf grows up to 16 inches tall and originates from a deep-seated, short rhizome. The few large leaves have blades about 6 inches long that are pinnately divided almost to midrid and extend beyond the round flower clusters. The stamens project much beyond the white, lavender, or purplish-blue corollas. The species is found higher up in the Boise Foothills in somewhat moist habitats.

Ecology & Ethnobotany Ballhead waterleaf blooms early in the spring, slightly after the more numerous glacier lily and yellow bells, but its herbage does not wither so quickly.

The young shoots, leaves, and flowers can be eaten raw, or these and the roots may be cooked and eaten. We find them exceptionally good in salads, or when eaten as a trail nibble. They do have a texture that takes some getting used to.

The leaves can be used as a protective dressing for minor wounds and are slightly astringent. As a poultice, it can be used for insect bites and other minor skin irritations.

BASIN NEMOPHILA
Nemophila breviflora   HYDROPHYLLACEAE

General Description This is a weak, angled, prostrate to ascending plant with stems up 8 inches tall. The plant is sparsely set with prickles. Leaves are fringed with stiff hairs, alternate, and pinnately cleft into 2 pairs of oblong to lance-shaped lobes. The corolla is bell-shaped, lavender colored, and not longer than the calyx lobes. The style is deeply bifid.

Ecology & Ethnobotany The genus name is from the Greek for "grove loving," referring to the woodland habitat of many species in this genus. The roots were used to prepare a decoction to cure asthma by some Native American tribes.

PHACELIA
Phacelia   HYDROPHYLLACEAE

General Description At least 4 species occur in the Boise Foothills. Phacelias include herbaceous annuals, biennials, and perennials with alternate leaves and various degrees of pubescence. The 5-parted flowers have

stamens extending beyond the corolla and are arranged in inflorescences that are spirally coiled and congested at first but elongate with maturity. Flowers bloom progressively from base to tip. The genus is challenging to work with in the field and the species sometimes hybridize.

Ecology & Ethnobotany As with many species in the Boraginaceae, contact with the hairs of some species of *Phacelia* can cause a very unpleasant rash similar to that from poison ivy in sensitive individuals.

ROCKY MOUNTAIN IRIS
Iris missouriensis   IRIDACEAE

General Description The stems of this plant are about two feet tall with several grass-like leaves. The flowers are showy with three drooping blue sepals and three petals that are slightly smaller than the sepals. Rocky Mountain iris can be found in meadows, wet or moist areas at low to mid elevations.

Ecology & Ethnobotany Members of this genus contain irisin, an acrid resin concentrated mainly in the rhizomes, and present in the foliage and flowers. People who raise irises sometimes develop a skin rash from handling the rhizomes. The leaves can be used to make crude cordage in making nets and snares.

DOUGLAS' GRASSWIDOW
Olsynium douglasii   IRIDACEAE

General Description The stems arise singly or in small clusters, somewhat flattened in cross-section, and bearing several simple, parallel-veined leaves. The leaves are as tall as the stem, or often shorter. The one to three flowers are deep purplish-red to occasionally white or white with purplish stripes. The six tepals are regular (same shape), expanding to greater than the diameter of a quarter. They are rounded at the tip. The

3 yellow tipped stamens are shorter than the elongated style, which is 3-pronged at the tip. The filament tube is only slightly enlarged above the base.

Ecology & Ethnobotany This genus is very similar to *Sisyrinchium*, but the filaments are united only at the base and the flowers are never blue. This species is found in dry open areas which are seasonally wet during the early spring.

There is only one species in the West, with two varieties. The variety *inflatum*, from mostly east of the Cascade Range, has a distinctly flared stamen base. The variety *douglasii* has an indistinctly flared stamen base; it occurs mostly west of the Cascade Range, from British Columbia south to northern California.

HORSE MINT
Agastache urticifolia LAMIACEAE

General Description This is a tall perennial growing 3 to 6 feet tall. The leaves are opposite, ovate in shape, 1 to 3 inches long and $1\frac{1}{2}$ inches wide. They are also coarsely toothed on the margins. Flowers occur in dense whorls and form a terminal spike, $1\frac{1}{2}$ to 6 inches long. The calyx is green or rose, and the corolla is 2 lipped, rose or violet, and 3/8 to 5/8-inch long. In the Boise Foothills, this plant is found in moist places at the higher elevations.

Ecology & Ethnobotany The seeds of horse mint may be eaten raw or cooked, and the leaves can be used in tea or for flavoring stews. Some Native Americans drank an infusion made from the leaves to relieve rheumatic pain, and for indigestion and stomach pains. The plant is said to have mild sedative qualities. The mashed leaves were made into a poultice and applied to swellings.

GROUND IVY
Glechoma hederacea LAMIACEAE

General Description This is a "rough to the touch" non-native perennial plant. The leaves are heart-to kidney-shaped, with rounded teeth. The flowers are violet-blue and occur in the leaf axils. It is native to Eurasia and has become established across the United

States. It is found in disturbed and moist habitats at the lower elevations. Unlike other mints, ground ivy does not have any discernible minty odor.

Ecology & Ethnobotany The plant is said to possess astringent and expectorant properties. Folk uses of the plant include a poultice that was used topically for treating bruises and muscle aches. It was also said to have been used in combating a variety of respiratory ailments, including bronchitis, pneumonia, and coughs.

HENBIT DEADNETTLE
Lamium amplexicaule   **LAMIACEAE**

General Description This is a weedy annual from Eurasia. The leaves are rounded to heart-shaped with shallow, rounded teeth on the margins. The flowers are pinkish-purple arising from the leaf-like bracts. It is found in waste places and fields within the Boise Foothills.

Ecology & Ethnobotany The entire plant is edible. We found it best when added to other salad plants. The plant is mildly astringent and a tea from the

leaf and flower is used by herbalists for minor internal or external bleeding, to relieve diarrhea, and other digestive problems. *Caution* - In most cases the plant may be found in areas that are often sprayed with herbicides and pesticides.

WATER HOREHOUND
Lycopus americanus   LAMIACEAE

General Description This plant grows up to 3 feet tall. It has smooth foliage and short-petioled lanceolate leaves that are toothed. Flowers occur in axillary clusters. The species is found in moist habitats. Unlike other mints, bugleweeds do not have the mint-like odor.

Ecology & Ethnobotany The leaves of American water horehound are edible raw but are usually tough and bitter.

WHITE HOREHOUND
Marrubium vulgare   LAMIACEAE

General Description This is a woolly perennial herb with bitter sap. The leaves are wrinkled and toothed. The flowers are small, white and occur in dense whorls. Horehound is a common weed of waste places and fields at the lower elevations.

Ecology & Ethnobotany The most famous use of this plant is horehound candy and is used to soothe sore throats and coughs. A tea from the dried leaves and flowers is also used, but because of the extreme bitterness of the herb, it is obvious why it tastes better in the form of a candy.

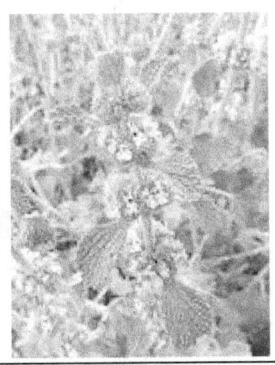

WILD MINT, SPEARMINT
Mentha   LAMIACEAE

      General Description These are distinctly aromatic perennial herbs with rhizomes. The flowers are arranged in whorls. The species are generally found in moist habitats in the Boise Foothills.

      Ecology & Ethnobotany The fresh or dried leaves of *M. arvensis* (wild mint), *M. canadensis* (Canadian mint), and *M. spicata* (spearmint) can be steeped in hot water for a tea. They have also been used as flavoring agents for soups, meat, and pemmican. The young leaves can also be added to salads and soups. The plants are high in Vitamin A, C, K, and minerals iron, calcium and manganese. It is an appetite stimulant and digestive aid.

CATNIP
Nepeta cataria   LAMIACEAE

      General Description This is a tap-rooted perennial that feels like felt to the touch. Leaves are triangular-shaped and coarsely toothed. The flowers are blue or yellowish-white in terminal, spike-like inflorescences. This introduced species from Europe is now widespread across North America and occurs in

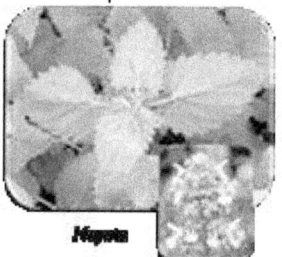

waste and disturbed places, and along irrigation canals.

      Ecology & Ethnobotany The nutritious young leaves and buds can be added to salads. The dried leaves and flowers make an excellent tea and are high in trace minerals and vitamins. The tea is a subtle, relaxing sedative on humans. As with other mint teas, it is also a carminative, and is soothing to an upset stomach.

SELF-HEAL
Prunella vulgaris   LAMIACEAE

General Description This perennial grows 4 to 20 inches tall, and has opposite, lanceolate ovate shaped leaves that are 1to 2 inches long. The herbage is glabrous to short pubescent. Flowers occur in a dense, terminal spike in the axils of round, membranous, purple tinged bracts. The sepals are purplish and the petals are 2-lipped and violet. Self-heal is common in moist places in the Boise Foothills and flowers from May to September.
Ecology & Ethnobotany The entire plant is edible, raw or cooked. However, we found that it is the young and tender plants collected in the early spring that are best. The crushed leaves can be used fresh or dried to make a tea.

PURPLE SAGE
Salvia dorrii   LAMIACEAE

General Description Members of this genus can be annual or perennial herbs or shrubs. This species grows up to 24 inches tall. The foliage is scurfy and grayish. Flowers are bluish or (rarely) white, in dense clusters near tips of stems. Stamens projecting well beyond corollas. Tends to grow in dry, sagebrush habitats in the Boise Foothills.
Ecology & Ethnobotany The seeds of perhaps all species of *Salvia* may be eaten raw or parched and

ground into meal. The seeds can also be soaked in water for a flavorful drink. Leaves of any fragrant sage can be used as a tea or spice for soups and meats. Sage does contain moderate amounts of Vitamin A and C, and can be added fresh to salads and sandwiches, however, we advise you to do this sparingly.

NARROWLEAF SKULLCAP
Scutellaria angustifolia   **LAMIACEAE**

General Description This plant grows up to 20 inches tall. Flowers are bluish violet in color. Flowers are axillary or occur in terminal racemes, or they may be solitary in the axils of leaves. Calyx is bell-shaped and 2-lipped with a gibbous protuberance on the upper side, splitting to the base at maturity. Corolla with an elongated curved ascending tube and dilated at throat. There are 4 stamens, the lower pair longer. Skullcaps are found in wet or moist soils

Ecology & Ethnobotany All the species contain scutellarin, the primary active compound that has been confirmed to be a sedative and has antispasmodic qualities.

WILD ONION
Allium   **LILIACEAE**

General Description Several species of onions occur in the Boise Foothills. They all arise from bulbs and have the characteristic distinct onion odor. The small flowers are clustered together in umbels.

Ecology & Ethnobotany *Allium aaseae* (south Idaho onion) is a small, early spring-flowering perennial plant endemic to southwestern Idaho, and is restricted to the lower foothills, from Boise to Emmett. The species is threatened by mining, housing developments, invasion of weeds, off-road vehicle use, and trampling by domestic livestock.

All species are edible. The bulbs may be eaten raw, boiled, steamed, creamed, in soup, and are especially good when used as a seasoning. Ingestion of large amounts of onions, including the cultivated ones, can cause poisoning or cause goiter, but are otherwise not known to be harmful. The seeds and leaves can also be eaten. Onions will keep a long time, because the skin dries and preserve the flesh inside. Onions contain large amounts of some important micronutrients, more Vitamin C than an equal weight of oranges, and more than twice as much Vitamin A as an equal weight of spinach.

Medicinally, people have taken advantage of onions natural antiseptic properties by applying the juice to wounds to prevent infection. The juice was also used as an insect repellent when rubbed over the body. The onion smell apparently has some beneficial effects on the circulatory, digestive, and respiratory systems.

SEGO LILY, MARIPOSA LILY
Calochortus   **LILIACEAE**

General Description Three species of *Calochortus* can be found in the Boise Foothills. They are characterized as perennials from bulbs, with tulip-like flowers that are few and showy. These species can be found in dry open places.

Ecology & Ethnobotany With their grass-like leaves and stem and their exotic flowers, the various species of *Calochortus* are favorites among many wildflower enthusiasts. The genus *Calochortus* comes from the Greek *kalos* (beautiful) and *chortos* (grass) referring to the grass-like leaves. Mariposa is Spanish for butterfly and Sego is a Shoshonean word for edible bulb.

The whole plant is edible raw or cooked. The biggest threat to these plants is overharvesting by people. Consider these plants only in an emergency.

GLACIER LILY

Erythronium grandiflorum   LILIACEAE

General Description. This is the only species in the area. The bright yellow flowers hang from a slender stalk. The two flat leaves are sheathing at the base. Glacier lily can be found up to and above timberline, blooming at the edge of melting snowbanks.

Ecology & Ethnobotany These beautiful flowers are seldom abundant, so glacier lily *should only be considered in extreme emergencies*. Harvesting destroys the plant. The young plants can be boiled as a potherb. The seed pods can be eaten raw or cooked. Eating the seed pod will not destroy the plant as long as you spread the seeds around first. Corms may be eaten raw but are better if boiled for at least 20 minutes. They can also be dried and stored for future use. The corms contain inulin, which is inedible raw. Normally they would be pit cooked for an extended period of time.

SPOTTED FRITILLARY, YELLOW BELLS
Fritillaria atropurpurea & F. pudica   LILIACEAE

General Description These are glabrous, perennial herbs from bulbs with numerous white bulblets around the base. The flowers are usually nodding with similar petals and sepals (tepals). Spotted fritillary (*F. atropurpurea*) flowers are brown or purple, mottled with yellow; yellow bells (*F. pudica*) flowers areyellow, fading to red and is one of the earliest of bloomers in the upper Boise Foothills.

Comparison of *Fritilaria* species

F. pudica Flowers yellow, fading to red
F. atropurpurea Flowers brown or purple, mottled with yellow

Ecology & Ethnobotany The bulbs of this genus have been a staple for Native Americans since prehistoric times. Bulbs of all species are edible raw or cooked but are relatively rare and should be considered only in an emergency.

FALSE-LILY-OF-THE-VALLEY
Maianthemum   LILIACEAE

General Description These are annual herbs with extensive, horizontal rootstalks. Leaves are alternate and sessile or on short petioles. Two species can be encountered in the upper Boise Foothills in moist or shady areas.

Comparison of *Maianthemum* species

Feathery false-lily-of-the-valley (*M. racemosum*) Perianth segments, less than 1/8-inch long; fruit is a red berry

Starry false-lily-of-the-valley (*M. stellatum*) Perianth segments ¼ inch long; fruit a purple to black-colored berry

Ecology & Ethnobotany Both species have edible berries that are not especially palatable. If eaten in quantity they can act as a laxative (hence the other common name of scooter berry). Cooking the berries removes much of the purgative elements making them a bit more palatable.  They are high in Vitamin C. The young shoots and leaves can be used like asparagus or eaten as a potherb. False-lily-of-the-valley have starchy rootstocks that may be eaten. However, the rootstocks must be soaked overnight in lye. Native Americans in Canada, used the white ash from their fire pits instead of lye, which supposedly removed the bitterness. The roots are then boiled and rinsed several times to remove the lye.

The mashed rootstock of starry false-lily-of-the-valley was thrown into a stream as a fish stupifier, making the fish easier to catch.

ROUGHFRUIT FAIRYBELLS
Prosartes trachycarpa   LILIACEAE

General Description This species was once included in the genus *Disporum*. This species has crisp-hairy stems and egg-shaped to broadly lance-shaped leaves. They are rounded at the base with spreading hairs along the margins. The creamy white tepals are shorter than the stamens, and the style is 3-parted at the tip. The nearly round berry is densely covered with minute bumps. In the upper portions of the Boise Foothills may be found in moist to somewhat dry habitats.

Ecology & Ethnobotany *Prosartes* is Greek and means *fastened* and somehow refers to the manner in which the fruit parts are attached. *Trachycarpa* is also Greek and means *rough fruit*.

As noted previously, this plant was formerly known as *Disporum trachycarpum*. Rodents and grouse eat the berries. For humans too, the berries are edible and mildly sweet, but mostly mealy and bland.

A poultice of the leaves can serve as a bandage for bleeding wounds, and an infusion can be used as a wash for wounds.

CLASPLEAF TWISTED-STALK
Streptopus amplexifolius   LILIACEAE

General Description Twisted stalk is a perennial herb with creeping rootstocks. The sessile or clasping leaves are alternate, elliptical to ovate in shape, and the flowers are yellowish-green. The 1-2 pendant flowers hang from the axils of the upper leaves on stalks that are bent in the middle. In the Boise Foothills I have encountered this species along some newly created bike trails around Bogus Basin where the trail cut through willow thickets.

Ecology & Ethnobotany In terms of edibility,  these plants have escaped mention in many guides but are indeed safe. The new spring shoots and clasping young leaves can be eaten raw or added to salads and taste somewhat like cucumbers. The berries, often referred to as watermelon berries are somewhat laxative if eaten in excess but may be eaten raw or cooked in soups and stews. They are sometimes referred to as "scooter berries," because if you eat too much you can find yourself "scooting" off to the bathroom. The species are easy to grow in wild gardens. The stems were used in poultices for cuts.

Warning Anyone wishing to use the young shoots of twisted stalk should be very careful to identify it correctly. At the shoot stage, these plants resemble the highly toxic corn lily (*Veratrum*).

DEATH CAMAS
Toxicoscordion **LILIACEAE**

General Description Previously included in the genus *Zigadenus*. At least three species may be encountered in te Boise Foothills. They are glabrous perennials with bulbs and grass-like leaves. The cream-colored to greenish white flowers are stalked and subtended by narrow bracts in an elongated inflorescence.

1a. Ovary slightly inferior; perianth segments about $\frac{1}{4}$ inch long, not clawed; glands of the perianth obcordate in shape ----- **mountain deathcamas (*T. elegans*)**
1b. Ovary wholly superior; perianth segments less than $\frac{1}{4}$ inch long; inner segment clawed glands not obcordate ----- **2a**

2a. Stamens distinctly surpassing the perianth; stem stout; inflorescence paniculate ----- **foothill deathcamas (*T. paniculatus*)**
2b. Stamens not much surpassing perianth; inflorescence usually a raceme ----- **meadow deathcamas (*T. venenosus*)**

Ecology & Ethnobotany Death camas is very poisonous plant if ingested. The alkaloids, primarily concentrated in the bulbs, can cause muscular weakness, slow heartbeat, subnormal temperature, stomach upset with pain, vomiting, and diarrhea, and excessive watering of the mouth. Death camas should not be confused with the edible camas (*Camassia*), which does occur in our area, but which formed a staple food for aboriginal peoples in the Pacific Northwest. It is also difficult to distinguish death camas from other edible plants, including wild onion (*Allium*), Mariposa lilies (*Calochortus*), fritillaries (*Fritillaria*), and brodiaeas (*Brodiaea*) prior to flowering.

LARGEFLOWER TRITELEIA
Triteleia grandiflora   LILIACEAE

General Description The flowering stems are erect and the leaves are few, basal, and grass-like. The leaves often wither away before the flowers appear. The flowers occur in an umbel and the segments all look alike. The fruit is a capsule. These plants occur on grassy and sagebrush slopes and flats within the Boise Foothills up into the forest.

Ecology & Ethnobotany The corms of most *Triteleia* species are edible raw but are somewhat mucilaginous. It is better if they are boiled for a few minutes or roasted. They can also be mashed and dried for future use in stews. Since the corms grow deep, it is usually easiest to harvest them with digging sticks.

CORN LILY, FALSE HELLEBORE
Veratrum californicum & V. viride   LILIACEAE

General Description These are characterized as large, coarse, leafy-stemmed herbs with thick rootstocks. The leaves are broad, clasping, and strongly veined. The numerous green to dull white flowers are arranged in a large panicled inflorescence. False hellebore is found in wet meadows and forest openings from the middle elevations into the alpine zone. Two species may be found in the upper Boise Foothills in wet meadows and along stream banks.

Comparison of *Veratrum* Species

V. californicum Lower panicle branches spreading to ascending; perianth mostly white
V. viride Lower panicle branches drooping; perianth green or yellow-green

Ecology & Ethnobotany These plants are very poisonous if ingested and have an inconsistent mixture of several powerful alkaloids. Some of the symptoms include depressed heart action, salivation, headache, burning sensation in the mouth, slowing of respiration, and death from asphyxia. These violent symptoms of poisoning may occur within 10 minutes. Avoid any use of the plant that involves ingestion. In some cases, just handling *Veratrum* can cause severe itchiness and irritation. Even nectar in the flowers is poisonous to insects and can cause serious losses among honeybees. **Personal observation** – I have observed many scout leaders mistakenly recommend their scouts to use the leaves to wrap food for pit cooking. *DO NOT FOLLOW THIS FOOLISH ADVISE!*

LEWIS' FLAX
Linum lewisii   LINACEAE

General Description This much branched annual has blue or rarely white flowers and alternate, sessile, and linear leaves. In the Boise Foothills this species grows on open slopes and flowers from May to July.

Ecology & Ethnobotany The seeds contain a cyanide compound but are edible after roasting them. They have a high oil content that contains essential fatty acids that are very much needed in our daily lives, plus they add an agreeable flavor to cooked foods. The stems are a source of linen, a fabric used for clothing.

BLAZINGSTAR
Mentzelia   LOASACEAE

      General Description The common name, blazingstar, actually describes only the tall and showy species. The little-known annuals are delicate and inconspicuous. The herbage of these plants is generally rough to the touch or with barbed, sometimes stinging hairs. The alternate, simple, entire to pinnately cleft and brittle leaves adhere to any foreign object contacted. Hence the other common name of stickseed. Flowers are bisexual, radially symmetrical, and borne singly or in a convex or flat-topped cluster.

      Ecology & Ethnobotany *Mentzelia* was considered an important food source in many places of the West. The seeds of most species may be parched, and then ground into flour. The seeds can be stored for future use. The Hopi Indians in the southwest parched and ground the small, oily seeds of *M. albicaulis* into a fine, sweet meal and ate it in pinches.

STREAMBANK WILD HOLLYHOCK
Iliamna rivularis   MALVACEAE

      General Description Hollyhock is a unique and stately perennial with 1-several stems up 5 feet tall. All green plant parts are set with star-shaped hairs. Leaf shape is variable with at least some reminiscent of 3- to 7-lobed maple leaves. The lavender to pink flowers are arranged on stout stalks in rather dense axillary or terminal clusters. Blooming occurs from June through August.

      Ecology & Ethnobotany This plant is often found along stream courses, in deep, moist but well-drained soil. It can also be found on disturbed sites such as clearcuts.

COMMON MALLOW
Malva neglecta   MALVACEAE

General Description The species is distinguished by its distinctive fruit and seeds, rather than their leaves and flowers. It is an introduced annual herb that is usually found in waste places at the lower elevations.

Ecology & Ethnobotany The entire plant of *M. neglecta* is edible. The young leaves are particularly good in salads or cooked up as a potherb. The plant is, however, very mucilaginous, and it is often used to thicken soup and may take a little getting used to. Eaten in large amounts, however, may cause digestive disorder. The immature fruits (which look like cheese) can also be eaten raw or added to soups.

GLOBEMALLOW
Sphaeralcea grossulariifolia & S. munroana
MALVACEAE

General Description These are perennial herbs that have star-shaped hairs on the leaves and stems. Flower colors range from red to pink. They can be found in open areas at the lower elevations. The genus name comes from the Greek *sphaira* and *alkea* to mean "spherical mallow."

Ecology & Ethnobotany Scarlet globemallow was chewed and applied to inflamed sores and wounds as a cooling, healing salve. It was also used as a pharmaceutical aid. The entire plant was ground and steeped in water for a sweet tasting tea that was mixed with other bad tasting medicines to make them more palatable. The Navaho Indians used the plant as a lotion to treat skin diseases, as a tonic to improve appetite, and as a medicine for rabies.

WOODLAND PINEDROPS
Pterospora andromedea
MONOTROPACEAE

General Description This is a brownish-red plant with sticky stems up to 3 feet tall with pale-yellow flowers. It can be found in deep humus of coniferous forests in the upper Boise Foothills and is usually associated with ponderosa pine (*Pinus ponderosa*). Flowers from June to August.

Ecology & Ethnobotany Some authorities prefer to call this plant a parasite rather than a saprophyte because it has no root system or fungi of its own. Instead, it draws its nutrition from mycorrhizae that are associated with adjacent plants.

Pinedrops has been used medicinally by some Native Americans. For example, a cold tea made from the pounded stems and fruits was used to treat bleeding from the lungs. As a dry powder, the plant was used as a snuff for nosebleeds. However, the plant does contain various poisonous compounds and should be avoided.

BOISE SAND-VERBENA
Abronia mellifera var. *pahoveorum*
NYCTAGINACEAE

General Description This is a genus of some 35 species of sprawling herbaceous annuals or perennials from coastal and desert areas of western North America. *Abronias* have branching, usually sticky-hairy stems, thick, toothless leaves occurring in pairs. *Abro* is Greek for delicate or pretty, referring to the flowers.

Ecology & Ethnobotany In the Boise Foothills one is likely to encounter this species in sandy substrate. The species is threatened by housing development, recreation, grazing, and invasion of non-native plants such as cheatgrass. The easternmost historical location is near Lucky Peak Dam east of Boise; sporadic populations occur (or occurred) northwest from here along the Boise Front to Horseshoe Bend, and west in the sandy ridge separating the Boise and Payette rivers (see Ertter, B. and S. Nosratinia. 2016. A new variety of *Abronia mellifera* (Nyctaginaceae) of conservation concern in southwestern Idaho. Phytoneuron 2016-20: 1–4. Published 3 March 2016.)

EVENING-PRIMROSE
Camissonia   ONAGRACEAE

General Description These are annual plants with basal or alternate leaves. The inflorescences are bracted and nodding, and the four sepals are reflexed. The white or yellow flowers usually fade to red.

Ecology & Ethnobotany Members of this genus were formerly included in the genus *Oenothera*; in summary, the species of *Camissonia* are distinguished by having a club- or head-shaped stigma, instead of the 4-part-divided stigma seen in *Oenothera* or *Clarkia*.

FIREWEED
Chamerion angustifolium   ONAGRACEAE

General Description Previously in the genus *Epilobium*. This is large perennial growing up to 5 feet tall. The leaves are alternate, lanceolate in shape up to 8 inches long. The edges of the leaves are more or less toothless, green colored above and pale beneath. Flowers occur in a long raceme. The 4 sepals are lavender-tinged and the petals are rose-purple in color. There are 8 stamens. Look for fireweed in the forest habitats of the Boise Foothills,

Ecology & Ethnobotany Fireweed is one of the "MUST KNOW" plants when it comes to survival. Food, drink, tinder, twine, and medicine are all provided by these abundant herbs.

In general, they are survivors in landscapes that have been ravaged by manmade and natural forces (e.g., fires, clearcuts). Soil conditions do appear to affect their flavor.

The most distinctive identifying feature of fireweed is the unique leaf venation. Unlike other plants, the veins do not terminate at the edges of the leaves, but rather join together in loops inside the outer margins.

SMALL ENCHANTER'S NIGHTSHADE
Circaea alpina   ONAGRACEAE

General Description Enchanter's nightshade is a delicate perennial growing up to 12 inches tall, with glandular hairs in the upper portions and very slender rootstocks. Heart-shaped leaves are 1 to 3 inches long, with subentire to sharply toothed margins and narrow, pointed tips. The flowers are inconspicuous with tiny, 2-petaled corollas.

Ecology & Ethnobotany A circumboreal species, it is most often associated with cool, moist and shaded sites. Despite the name, it is not an alpine species; rather it occurs predominantly in montane and lower subalpine

PINKFAIRIES, DIAMOND CLARKIA
Clarkia pulchella & C. rhomboidea   ONAGRACEAE

General Description These are annuals with brittle stems and purple or red, showy flowers. In the Boise Foothills they are usually found on dry slopes up into the forest.

Comparison of *Clarkia* species

C. pulchella Petals 3-lobed at tip; leaves alternate
C. rhomboidea Petals not lobed at tip; leaves subopposite

Ecology & Ethnobotany The genus name honors Captain William Clark of the Lewis and Clark Expeditions to the Northwest in 1806. Seeds of this genus were collected by Native Americans.

After drying and parching them, the seeds were ground up into flour.

WILLOWHERB
Epilobium   ONAGRACEAE

General Description Six or seven species of willowherb may be encountered in the Boise Foothills. The genus includes annual and perennial plants that have willow-like leaves. The flowers are white or lavender in color with petals that are often notched. Fruits are long, narrow pods that open by 4 slits to release the numerous small, densely hairy seeds. The roots and pods are often needed to make positive identification of the many species.

Ecology & Ethnobotany The genus name is from the Greek, meaning "on a pod," describing the elongated ovary bearing the other flower parts on its top. The common name refers to the tufts of hairs at the end of the seed, which is similar to that on willow seeds.

GROUNDSMOKE
Gayophytum   ONAGRACEAE

General Description Plants in this this genus are comprised of inconspicuous, fragile, weedy-looking, small annuals. Stems are freely branched and slender with lance-shaped to linear, alternate, and entire leaves. The small flowers have distinct, reflexed sepals, white to pink petals, and 8 stamens. Fruits are linear to club-shaped capsules.

<u>Comparison of *Gayophytum* species</u>

Spreading groundsmoke (*G. diffusum*) & blackfoot groundsmoke (*G. racemosum*) These two species have stalked capsules. They are branched above, and the internodes are longer than the leaves.

Pinyon groundsmoke (*G. ramosissimum*) This species is branched from the base and has unstalked capsules and leaves longer than the internodes.

Ecology & Ethnobotany An infusion of *G. ramosissimum* was used to soothe irritated skin.

EVENING-PRIMROSE
Oenothera   ONAGRACEAE

General Description There are many species in this genus which are annual, biennial, and perennial herbs. The flowers are white or yellow, often opening at night. There are 8 stamens, 4 petals, 4 sepals, and the stigma is globe-shaped to deeply four-lobed. The various species can be found in a variety of habitats in the Boise Foothills.
Ecology & Ethnobotany It has been suggested that all species would stand a trial as none are known to be poisonous. The various species are known to hybridize

 easily making identification at times challenging. We have cooked and eaten the young seed pods of several species and found them to have an acceptable taste. Other experts we've

consulted with also suggest that many species have seeds that are edible after being parched or ground into meal.

REIN ORCHID

Piperia unalascensis   ORCHIDACEAE

General Description This orchid has leafless stems with narrow bracts above the base and 1-3 fleshy, egg-shaped roots. The 2-4 basal leaves have a sheathing base and a lance- shaped blade. Numerous small, green to yellowish flowers are borne in a long dense to open spike-like inflorescence. The upper sepal and lip petal are egg-shaped, while the upper petals and lower sepals are narrower and spreading. The cylindrical spur is curved and longer than the lip.

Ecology & Ethnobotany This species may be found in the forested portions of the Boise Foothills. The species in this genus were formerly placed in the genus *Habenaria*. The genus name honors Charles Vancouver Piper (1867-1926), an agronomist with the U.S. Department of Agriculture and an expert on Pacific Northwest flora.

The bulbs of *P. unalascensis* were baked and eaten like baked potatoes by the Pomo and Kashaya.

BROOM-RAPE

Orobanche fasciculata & O. uniflora
OROBANCHACEAE

General Description Two species occur in the Boise Foothills and parasitize the roots of other plants. These fleshy annual plants are nearly white to brownish or purplish in color and lack chlorophyll. The leaves are

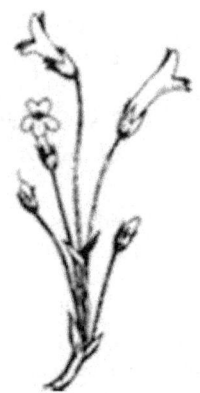

reduced to scales. Broom-rape is usually found in dry soils, associated with such genera as *Artemisia* (sagebrush) and *Eriogonum* (buckwheat).

Comparison of *Orobanche* species

O. fasciculata Flowers more than 3; the top of stems emerging from ground; petals pointed; sepals triangular
O. uniflora Flowers no more than 2-3, usually 1; the short stems remaining underground; petals rounded; sepals narrow and slender

Ecology & Ethnobotany The entire plant of broom-rape, roots and all, can be eaten raw. Being succulent plants, they answer for food and drink, and are often called sand food. We found them to be better tasting when roasted in the hot ashes of a campfire.

BROWN PEONY
Paeonia brownii   PAEONIACEAE

General Description The petals are reddish purplish with greenish-yellow borders, and the sepals are green. There are numerous stamens with golden anthers. In the center of the flower are 5 egg-shaped ovaries with no apparent styles but with a little cream-colored stigma. Leaves are light gray green. Look for this species in sagebrush and ponderosa pine habitats.

Ecology & Ethnobotany The genus name is from the Greek for Paeon, the physician of the gods who supposedly used the plant medicinally. *Paeonia brownia* is named for Robert Brown, an English botanist who lived from 1773-1858.

PONDEROSA PINE
Pinus ponderosa   PINACEAE

General Description Ponderosa pine is the most widely distributed conifer of western America. It has two or three long, dark-green needles per bundle; medium-sized cones that are reddish-brown when mature; and brown or blackish bark turning a distinctive orange color on older trees. They shed their seed the second year. Ponderosa pine may live to old ages. It often lives for 300 to 500 years and in a favorable environment forms nearly pure stands of open park-like forests. It is commercially valuable as wood for construction as well as interior finish.

Ecology & Ethnobotany All pines have edible seeds. However, they are an erratic food source, yielding an abundant crop in some years and a sparse crop in others. To collect the cones, long poles were used to knock them from the branches. One of the best ways to gather seeds is to heat the green cones until they open. The seeds are best when harvested in fall or early winter when cones normally release their seeds. The nutritious

seeds can then be shelled and eaten, or ground or roasted and made into flour. Seeds may contain as much as 15% protein, 62% fat, and 18% carbohydrates, with approximately 3,000 calories per pound.

The inner bark is also edible in an emergency. Though tedious, the tender mucilaginous layer between the bark and wood was scraped or peeled off. It was then cooked or ground into meal.

The firm and unexpanded pollen cones can be boiled and eaten. They have a surprisingly sweet and non-pitchy taste.

The needles of most pines can be steeped in hot water to make a satisfying tea and are a good source of Vitamin C. It also takes some practice to steep the right amount of leaves, since too much may be too strong. Additionally, the pine cleaning fluid can be extracted from boiling the needles and skimming off the oil-like substance from the surface. It may take a lot of pine needles to get a small cupful.

Pine sap can be collected in quantity from cuts, burns, and broken branches. The collected sap is then heated and formed into balls for future use. Be careful not to expose the sap to flames as it is very flammable.

DOUGLAS-FIR
Pseudotsuga menziesii   PINACEAE

General Description This is a tall tree, 35 to 60 feet tall with drooping branches. The leaves are needle-like, blue-green, and spirally arranged on the branches, but appear to be in a flat spray because the needles are turned at the petiole base. Needles are $\frac{3}{4}$ to $1\frac{1}{2}$ inches long and pointed at the tip. Cones are cylindrical, 4 to 6 inches long, with 3-fingered bracts overlapping the scales. These bracts are characteristic of the genus.

Ecology & Ethnobotany Douglas-fir is an important timber species. The wood is resinous with close, even, well-marked grains, and is of medium weight, strength, stiffness, and toughness. It is very durable, and when well-seasoned, does not warp. It is used in piles, ties, floors, and millwork, and to make a variety of items such as spear handles, spoons, fire tongs, and fishing hooks.

A tea can be made from the needles of Douglas-fir. Similar to pines, the pitch can be used as a glue. It can be used for sealing implements and caulking water containers. Medicinally, the sap provided a salve for

wounds and skin irritations. The pliable roots have been
used in weaving.

PLANTAIN

Plantago lanceolata & P. major   PLANTAGINACEAE

General Description These are characterized as short-stemmed annual or perennial herbs with basal leaves. The flowers are greenish or purplish. These two species are introduced from Europe and can be found at the lower elevations, particularly in fields and waste places.

Comparison of *Plantago* species

P. major Leaf blades broadly ovate, abruptly contracted to the petiole
P. lanceolata Leaf blades lanceolate, oblanceolate, lance-oblong, or elliptic, gradually tapering into the petiole

Ecology & Ethnobotany As a food, the young leaves of both species were used fresh or cooked. They contain calcium and other minerals. One hundred grams of plantain is said to furnish as much Vitamin A as a large carrot. The older leaves may be too fibrous and bitter for use, but they are usable if one is able to remove the fibers. Seeds are tedious to collect in quantity but can be ground and used as flour substitute or extender.

CRESTED WHEATGRASS
Agropyron cristatum   POACEAE

General Description This is a rhizomatous or bunch-forming perennial of mainly dry habitats. The leaves have flat or rolled blades and sheaths with free margins and usually evident auricles. The 3- to 12-flowered spikelets are sessile in a narrow, terminal spike. The 2 glumes in each spikelet are about equal in length and shorter than the lowest lemma. Both glumes and lemmas are sometimes awned from the tip.

Ecology & Ethnobotany Crested wheatgrass is a species originally introduced into the Intermountain region to improve degraded pasture and rangelands. It is now considered a problematic weed, capable of forming large, monospecific stands that exclude native vegetation and reduce forage diversity.

All members of this genus are useful. The plants often grow in abundance and the seeds can be collected and ground into flour. With some species, there may be long roots which can be been dried and ground into flour.

BENTGRASS
Agrostis   POACEAE

General Description The bentgrasses are tufted or rhizomatous annuals or perennials with sheaths that have free margins. Spikelets are stalked and borne in an open or contracted inflorescence. Each spikelet has only 1 floret. The glumes are about equally long and mostly pointed at the tip. The lemma is shorter than the glumes and has long to short hairs at the base and sometimes an awn arising from the back. The palea is very small or lacking.

Ecology & Ethnobotany Another edible group of grasses. The seeds can be made into flour.

WILD OATS
Avena   POACEAE

General Description These are introduced annuals with flat leaf blades and sheaths that have free margins. The spikelets are 2- to 3-flowered and borne on flexible stalks in an open, branched inflorescence. The glumes are longer than the lowest lemma. The lemmas are leathery with a bent awn arising from near the middle.

Ecology & Ethnobotany The seeds are edible. The hairs must be singed off first and then the grains can be ground and used as flour. The seeds are approximately 15% protein and 11% fat.

BROME GRASS
Bromus   POACEAE

General Description Bromes are native or introduced, annual or perennial, usually tufted grasses. The leaves have flat blades and glabrous or hairy sheaths with joined margins. The several-flowered, stalked spikelets are flattened and borne in an open to contracted inflorescence with nearly erect to drooping branches. The pointed or blunt-tipped glumes are of unequal lengths and shorter than the lowest florets. The lemmas are often awned from the tip.
Some of the introduced annual species have come to dominate disturbed grasslands throughout the semi- arid west.

Ecology & Ethnobotany The seeds of perhaps all species are edible. They can be gathered and cooked into a gruel.

Tongue in cheek humor - cheatgrass
(*Bromus tectorum*) - Idaho's State flower.

On the serious side of things, cheatgrass is a native plant in parts of Africa, Asia, and Europe and began to arrive in the New World sometime in the 18th century. It was initially present in small numbers, but it has since become one of the most widely distributed plants (invasive or native) in all of North America. The history of cheatgrass is tied closely to the history of western expansion. As the western United States was grazed by livestock, plowed under for cultivation or highway construction, cheatgrass took advantage of these disturbances.

Because cheatgrass is a winter annual species. It emerges in fall, goes dormant in winter, and becomes active again when spring arrives. Cheatgrass grows rapidly in the spring, and generally produces seed and dies by May or June. By completing its life cycle so early in the growing season, cheatgrass can be very competitive against native plants.

ANNUAL HAIRGRASS
Deschampsia danthonioides   POACEAE

General Description These are tufted perennial or annual grasses with flat or rolled blades and leaf sheaths that are open along the margins. The spikelets are mostly 2-flowered and borne on spreading to erect branches of narrow or open inflorescences. Glumes are mostly of unequal length and longer than the lower floret. The lemmas are toothed at the tip and have long hairs at the base and a straight or bent awn attached about the middle of the back.

Ecology & Ethnobotany Hairgrass is often used for wetland restoration, erosion control, and revegetation of other moist, disturbed areas where quick, low growing ground cover is desired. Waterfowl and birds eat the seeds. The seeds can be used in making flour.

WILD RYE, SQUIRRELTAIL GRASS
Elymus elymoides   POACEAE

General Description The ryegrasses are bunch-forming or rhizomatous perennials with hollow culms and leaf sheaths with free margins. The inflorescence is a terminal spike with 2 spikelets per node. The spikelets have 2-6 florets, and the narrow glumes sometimes have a short awn and are nearly equal in length. The lemmas are rounded on the back, with or without awns.

Ecology & Ethnobotany Squirreltail is considered to be one of the most fire resistant native bunchgrasses. This species is adapted to a wide range of ecological conditions and plants can be found from 2,000 to over 11,500 feet in elevation in desert shrub to alpine communities. Probably any *Elymus* produces seeds which can be used as food.

FESCUE GRASSES
Festuca   POACEAE

General Description The members of this group of genera are annual or perennial, rhizomatous or bunch-forming grasses with hollow culms. The leaves have flat, folded, or rolled blades and sheaths with free margins. The 2- to 12- flowered spikelets are borne on erect to drooping branches of the open to contracted inflorescence. The narrow glumes have pointed tips and are usually unequal in length and shorter than the lemmas.

The lemmas are rounded on the back and awnless or with a short awn-tip.

Ecology & Ethnobotany The seeds of probably all species are a food item.

FOXTAIL BARLEY
Hordeum   POACEAE

General Description Our wild barleys are tufted annuals or perennials with flat leaf blades and sheaths with free margins. The mostly 1-flowered, sessile or short-stalked spikelets are borne, usually 3 per node, in a dense, terminal spike. At each node, the central spikelet has both stamens and a pistil, while the lateral spikelets only have staminate or sterile florets. The glumes are narrow and awnlike. The lemmas are rounded on the back, usually with a long awn.

Ecology & Ethnobotany Seeds were collected, ground into flour, and used as food. In India, a cooling drink called sattu is made from barley.

Barley (*H. vulgare*) is the fourth most important cereal in the United States.It is used for livestock fodder, and beer brewing industry.

BLUE GRASS
Poa   POACEAE

General Description These are native or introduced, annual or perennial, bunch-forming or rhizomatous grasses. The leaves have sheaths with free margins and flat, folded, or rolled blades, usually with tips shaped like the bow of a boat. The 2- to 7-flowered spikelets are borne on nearly erect to spreading or reflexed branches of the open or contracted inflorescence. The glumes are equal or unequal in length and usually much shorter than the lemmas. The lemmas are variously hairy, unawned, and often prominently 5-nerved.

Ecology & Ethnobotany This is the largest genus of grasses in our area. Many of them are similar and difficult to tell apart without examination under a microscope.

Gather the seeds for food. As with many grasses, they do not remain on the plant for long, so the season for picking them is short.

INDIAN RICE GRASS
Stipa hymenoides   POACEAE

General Description The species may be known by several scientific names including *Oryzopsis hymenoides* and *Achnatherum hymenoides*. All is good as new research into genetics try to better understand how this and other plants are related to each other.

Ecology & Ethnobotany It is also an important native forage grass in western desert environments. This and other species of *Stipa* produce edible seeds that can be eaten raw but are best when dried and ground into flour. Use the flour to make cakes and mush.

COLLOMIA
Collomia grandiflora & *C. linearis*
POLEMONIACEAE

General Description Plants of this genus are annuals or perennials with funnel-shaped or tubular corollas with the throats that abruptly flare into an expanded limb. Two species occur in the Boise Foothills at the higher elevations.

Comparison of *Collomia* species

C. grandiflora Flowers salmon color to pale yellow, $\frac{1}{2}$ to $\frac{3}{4}$ inch long
C. linearis Flowers pink or lilac, flowers less than $\frac{1}{2}$ inch long

Ecology & Ethnobotany The genus name comes from the Greek *kolla* (glue), a reference to the seeds, which become gelatinous in texture when wet. The seed coat becomes mucilaginous when wet. This "glue" helps keep the germinated seeds from drying out. This is a mechanism that helps store water between the first autumn rains and those that may not come for several weeks.

From the roots of *C. grandiflora* an infusion was made to treat high fevers. Additionally, an infusion of the leaves and stalks was taken for constipation and to "clean out the system." *C. linearis* was used as a dermatological aid by the Gosiute. They used a poultice of the mashed plant and applied it to wounds and bruises.

SMALLFLOWER GYMNOSTERIS
Gymnosteris parvula POLEMONIACEAE

General Description This is a very small annual herb growing a thin green to red-colored stem only an inch or two tall. There are no stem leaves, though sometimes the persistent cotyledons may be visible. At the end of the stem is a large, fleshy inflorescence with red-tinged green bracts that serve as leaves. Within the inflorescence are one to five small flowers, each less than a ½ inch long. The yellow-throated flower has yellow or white oval-shaped lobes with pointed tips.

Ecology & Ethnobotany Gymnosteris is derived from the Greek for *naked stem*. It is self-pollinating. The species may be found on open, dry slopes and meadows from the foothills to fairly high elevations in the mountains. Often found growing under sagebrush and as such can be adversely affected by livestock grazing.

SCARLET GILIA
Ipomopsis aggregata POLEMONIACEAE

General Description This is a biennial plant growing 1-3 feet tall. The tubular or funnel-form flowers are red, orange, pink, or white and showy. The plant is usually found on dry slopes up to the subalpine. Within the Boise Foothill, scarlet gilia grows in a variety of habitats from desert canyons and cliffs to montane meadows, and subalpine rock fields. In short, in the upper portions of the foothills proper.

Ballhead ipomopsis (*I. congesta*) is also found and the two species may be distinguished in the following way:

Comparison of *Ipomopsis* species

I. *aggregata* Corolla tubes $\frac{3}{4}$ to 2 inches long
I. *congesta* Corolla tubes not over $\frac{1}{2}$ inch long

Ecology & Ethnobotany Scarlet gilia has also been called skunk flower because of a faint skunk-like smell from its glandular foliage. The plant was used by Native Americans as a tea to treat colds, to make glue, and to treat blood troubles. In Nevada, the principal use of this plant was for the treatment of venereal diseases. The whole plant was boiled for the purpose and a solution was taken as a tea or used as a wash. The whole plant was also boiled by the Ute Indians in Utah to make glue. A blue dye can be extracted from the roots.

GRANITE PRICKLY PHLOX
Linanthus pungens   POLEMONIACEAE

General Description This is a semi-evergreen, compact low-growing shrub that has several to many branching and densely leafy stems. Its distinctive prickly quality is due to the sessile, alternate and palmately divided leaves. The tips of each of the 3-7, stiff segments are set with a minute spine. The leaves usually persist for 1-several years. White or yellowish flowers have a long corolla tube with rounded lobes and a phlox-like fragrance. They are solitary in the axils of upper leaves.

Ecology & Ethnobotany The species was formerly included in the genus *Leptodactylon*. Granite prickly phlox was used as a decoction to bathe swellings, sore eyes, and scorpion stings.

SLENDER PHLOX
Microsteris gracilis   POLEMONIACEAE

General Description Formerly known as *Phlox gracilis*, this small branched annual grows 4 to 8 inches high and is glandular pubescent above. Leaves are entire, and the lower ones are opposite, while the upper ones are alternate. The leaves are oblong lanceolate. Flowers are small, usually in pairs in the upper leaf axils. The corolla is salverform, with yellow tube and white to purplish pink lobes. In the Boise Foothills, look for this species in grassy areas. It flowers from March to August.

Ecology & Ethnobotany This is a very common and very variable species. Slender phlox was eaten by some Native Americans as greens.

PHLOX
Phlox   POLEMONIACEAE

General Description These plants are dwarfed and mostly woody at the base and are somewhat sweet-scented. The corollas are distinctly tubular and expanded at the top into a lid (ring of petal lobes). The leaves are narrow, sessile, and opposite. This is a very difficult genus with many species that can only be distinguished with use of a hand lens. There are several species in the Boise Foothills and Forest.

Ecology & Ethnobotany A decoction from the roots of *P. longifolia* and other species were used by some Native Americans as eyewash for sore eyes. The scraped roots were soaked in water or steeped or boiled to make the wash.

AMERICAN BISTORT

Bistorta bistortoides   POLYGONACEAE

General   Description
This is a slender perennial with
glabrous stems. The leaves occur
mostly near the base and are
oblong lanceolate in shape with
sheathing stipules. The upper
leaves usually without petioles.
Flowers occur in a thick, cylindric,
pink to white terminal spike. In
the upper Boise Foothills, bistort
is found in wet areas such as wet
meadows and along streams.

Ecology & Ethnobotany Like *Polygonum*
(knotweed and smartweed), the young greens are useful
in cooking, the roots have a pleasant taste, eaten raw or
cooked.

WILD BUCKWHEAT

Eriogonum POLYGONACEAE

General Description There are a few species of
buckwheat in the Boise Foothills and Forest. In short,
they are annual or perennial herbs; some species are
woody at the base. The flowers are small and usually
bright colored.

One of the more common and obvious species is suphur buckwheat (*E. umbellatum*). It is a stoutly taprooted perennial with a freely branching crown and mostly prostrate branches that form mats up to 2 feet across. When the stems are ascending, they are up to 12 inches tall. Leaves are mostly less than 3 inches long, with the blade constituting about half or less of the length. The lower surface of the leaf is densely long-hairy and gray-white, and the upper surface is less hairy (smooth) and green. Blade shape is highly variable (mostly lance-shaped), but petioles are always slender. At the top of the tall flower stalk are several leafy bracts and 6 or more stalked, funnel-shaped involucres with long drooping

Lobes. The sulfur-yellow or off-white (pink-tinged) flower and its stalk are hairless.

Ecology & Ethnobotany The genus name is from the Greek *erion*, meaning wool, and *gony*, meaning knee or joint, referring to the hairy stems of many species. This is a complex genus with several intergrading and hybridizing taxa. The flowers are reduced and quite uniform.

None of the species are known to be poisonous. The flowering stems can be eaten raw or cooked before they have flowered. Seeds can be collected (though tedious) and ground into flour. A tea from the root of *Eriogonum* was used to treat headaches and stomach problems. The plants are mildly astringent and were used as a gargle for sore throats.

SMARTWEED, KNOTWEED
Persicaria & Polygonum   POLYGONACEAE

General Description There many species of knotweed in the Boise Foothills. In short, they are annual or perennial herbs with stems that are more or less swollen at the nodes. A sheathing stipule surrounds the stem at points where the leaves are attached. The mostly bisexual flowers are borne in axillary clusters or more compact inflorescences at the stem tips. The 4-6, mostly greenish, white, or pink perianth segments (tepals) are fused at least at their bases. The fruit, a lens-shaped to 3-angled achene, is usually completely enclosed by persistent perianth but may be extend well beyond it. Several species of *Polygonum* have in recent years been reclassified into the *Persicaria* and *Bistorta*.

Ecology & Ethnobotany Experimentation may be the rule for *Polygonum* (and the other genera mentioned) as none of the species are known to be poisonous. They do, however, vary in degrees of palatability. Tannins are found in the plants and large amounts might cause digestive upset and possible kidney damage. In moderate quantities, however, the genus is generally regarded as safe. Based on our experiments with various species, some have peppery tasting leaves that can be used in flavoring foods. Others have starchy roots that may be eaten raw or boiled and roasted. Still others have young foliage made into good salads or potherbs.

DOCK
Rumex   POLYGONACEAE

General Description There at at least 4 species of *Rumex* in the Boise Foothills. They are annual or perennial herbs with small flowers that are greenish and aggregated in a large terminal inflorescence. The fruits are called utricles. They can be found in many habitats. For fun, I've included a dichotomous key to the four *Rumex* species a person is likely to encounter in the area.

1a. Flowers nearly all unisexual and female and each sex borne on separate plants; leaves arrowhead-shaped or with basal lobes directed outward ----- **sheep sorrel (R. acetosella)**
1b. Flowers composed of both sexes; leaves of varied shapes, but not arrowhead-shaped ----- **2a**

2a. None of the valves (here, the inner perianth segment that encloses the achene) of fruit having hardened projection ----- **3a**
2b. All or part of valves of fruit having hardened projections ----- **veiny dock (R. venosus)**

3a. Stems unbranched below the inflorescence; petioles mostly with some degree of rough hairiness or with minute rounded projections ----- **curly dock (R. crispus)**
3b. Stems having axillary, often much reduced branches at some or all nodes below the inflorescence; petioles only rarely with hairs rough to the touch or with small bumps ----- **willow dock (R. salicifolius)**

Ecology & Ethnobotany The young leaves of dock can be used as greens and we have found that the flavor varies from species to species. The young leaves are best when collected before the flower stalk emerges. Also, because the leaves become watery when cooked, use very little water and don't overcook them. The older leaves may be too bitter for use. The leaves of dock are high in Vitamin C and contain more Vitamin A than carrots. Native Americans ground dock seeds and used the meal to make breads. However, removing the papery seed cover involves a lot of work, and depending on the species, is probably more work than it is worth. The distinctive sour taste of these plants is due to oxalic acid. As with other species that contain oxalic acid, docks should be used in small portions as they can cause calcium deficiency.

RED MAIDS

Calandrinia menziesii   PORTULACACEAE

General Description Formerly known as *Calandrinia ciliata*. This is a low glabrous annual plant growing up to 16 inches tall. The leaves are alternate, entire, linear to oblanceolate in shape and somewhat fleshy. The flowers are a magenta or red-violet color and there are 2 sepals and 5 petals. This is a fairly common species found growing in early spring on grassy slopes along the ICT.

   Ecology & Ethnobotany The seeds of all *Calandrinia* species are edible. After gathering and winnowing, the seeds should be parched with coals, pulverized, and then pressed into cakes for eating. The roots were also eaten, and the young leaves and stems were eaten raw or cooked.

BITTERROOT
Lewisia rediviva & L. sacajaweana
PORTULACACEAE

   General Description The genus is indigenous to the western parts of the U.S. and can be found clinging precariously to rocky ledges among boulders, on rock-strewn slopes, damp gravely places, alpine meadows, and in near desert conditions where rainfall is seasonal and unpredictable. Two species occur in the Boise Foothills and Forest; *rediviva* in the lower foothills and *sacajaweana* in the upper foothills along Boise Ridge Road. In the latter case, the author discovered several new population in 2016 and 2017 while conducting surveys.

   Ecology & Ethnobotany *L. rediviva* was an important food item for many Native Americans. The root is remarkably large and thick for a small plant and contains nutritious farinaceous matter that is much prized. The roots are dug up in spring before flowering. Once dug, the root is peeled promptly and the small red "heart" (embryo of next year's growth) is removed to reduce the roots bitter flavor. It is then steamed, boiled, or pit-cooked and eaten. The root can also be dried and will keep for a long time. The bitterness of the root varies and cooking is said to improve the flavor. The root boiled to a jellylike consistency will be pink in color. The pounded root was chewed for a sore throat.

Sacajawea's bitterroot (*L. sacajaweana* B.L. Wilson & E. Rey-Vizgirdas) is the first plant species to be named in honor of Sacajawea. This is an Idaho native and occurs nowhere else in the world but in the south-central Idaho mountains. Just over two dozen populations of Sacajawea's bitterroot are known to exist, roughly three-fourths of them on the Boise National Forest. Scattered  populations also occur on the Payette, Sawtooth, and Salmon-Challis National Forests.

In general, bitterroots are threatened by overgrazing and trampling by livestock and habitat destruction from road construction and recreation activities (e.g., trail construction). In the Boise Foothills, known populations are highly likely to be impacted by various activities including trail construction.

"SPRINGBEAUTY/MINERSLETTUCE"
Claytonia & Montia   PORTULACACEAE

General Description These are two closely related genera that may be best treated here together.

Claytonia - These are as perennial succulent herbs arising from deep corms or fleshy taproots. The leaves are opposite and the flowers white or pink in color.

Montia - This genus is comprised of slightly  succulent annual and perennial herbs. The flowers have two persistent sepals and five white or pinkish petals. Most *Montia* species grow in moist or seasonally wet areas that are partially to fully shaded. Formerly, several of the claytonias were classified as montias. I've included here a key from my 2017 book "*A FIELD GUIDE TO IDAHO MOUNTAIN PLANTS.*" Hopefully, you find in useful in identifying these two taxa.

1a. Plants with thick fleshy taproots or ovoid corms ----- *Claytonia*
1b. Plants with slender taproots, rhizomes, or stolons ---- - **2a**

2a. Stem leaves usually 2 (below the inflorescence), opposite or often connate and forming a disk that surrounds the stem at base of inflorescence ----- *Claytonia*
2b. Stem leaves more than 2, opposite or alternate ----- *Montia*

Ecology & Ethnobotany *Claytonia* - Often called Indian potato, wild potato, or mountain potato, the small corms can be eaten raw, boiled, or roasted. For many Native Americans, spring beauty was an important "root vegetable." When collecting, keep only the largest corms and replant the others. At first, many find the corms distasteful, as they do take a little getting used to. The corms are high in starch and, when cooked, taste like potatoes. Boil or bake the corm for 30 minutes. Most species are not plentiful, so be conservative in your endeavor. They can also be dried on strings for long-term storage. The rosettes can also be eaten raw or cooked and are high in Vitamins A and C. They are better when mixed with other salad plants.

Miner's lettuce (*C. perfoliata*) has stems and leaves that can be eaten raw or boiled like spinach. The roots are also edible raw or boiled. Some Native Americans picked miner's lettuce and placed it near the nests of red ants. The ants were allowed to crawl over the leaves and were then shaken off. The residue left on the leaves by the ants had an acerbic flavor.

All species of *Montia* have stems and leaves that can be eaten raw or boiled like spinach. The roots are also edible raw or boiled.

COMMON PURSLANE
Portulaca oleracea   PORTULACACEAE

      General Description Purslane is a smooth, succulent, ground-hugging annual whose prostrate stems form mats up to 2 feet in diameter. The leaves are inversely egg- to spatula-shaped and mostly 1-inch long. The small, greenish-yellow flowers are borne in small terminal clusters in the leaf axils. The conical fruits open transversely at their midpoint to release black, minutely pimpled seeds. This small succulent annual herb is found in disturbed habitats at the lower elevations of the Boise Foothills.

      Ecology & Ethnobotany The genus name may be derived from *portula* meaning "little gate," referring to the lid on the capsule. It has been used as a food for more than 2,000 years in India and Persia. In Europe, it is grown as a garden vegetable. It has been pickled, used to thicken soups, or dried or frozen for storage. Seeds are used in the form of flour.

BANEBERRY
Actaea rubra   RANUNCULACEAE

      General Description This is a perennial herb with fibrous roots. The leaves have long petioles and are 2-3 times divided into sharply toothed, lance-shaped segments. The small flowers are white and borne in a branched, congested, hemispheric inflorescence. The fruits are shiny red or white. In the Boise Foothills, baneberry will be found in moist, montane forests and riparian areas, usually with some partial shade.

      Ecology & Ethnobotany The ***entire plant, especially the berries, is poisonous***. The plant is sometimes confused with *Osmorhiza chilensis* (western sweetroot) which often shares the same habitat. However, unlike baneberry, sweetroot has a strong licorice-like odor.

For those of you having been scouts remember the adage "berries white poisonous site." Keep in mind baneberry produces both white and red berries and that in no way are the red berries edible. *I have met scout leaders with absolutely no experience in the outdoors and botany assume only that white berries are poisonous and that the red one are edible.*

COLUMBINE

Aquilegia formosa   RANUNCULACEAE

General Description This perennial herb with large divided leaves has red and yellow flowers that are nodding. Each of the 5 petals extends backward between the sepals forming a hollow spur. It is generally found in moist habitats, meadows and moist slopes in the upper Boise Foothills.

Ecology & Ethnobotany The flowers of columbine are edible and have a sweet taste but grow bitter with age. They can be added to salads in small amounts. **Warning** The seeds can be fatal if eaten and most parts of the columbine contain cyanogenic glycosides. Any therapeutic use of columbine is strongly discouraged.

BUR BUTTERCUP
Ceratocephala testiculata   RANUNCULACEAE

General Description This small annual has hairy foliage and ubranched, erect, leafless stems 1-3 inches tall. The basal leaves have slightly winged petioles and blades that are divided into 3-5 broadly linear segments. The solitary flowers have 5 petals and slightly smaller, persistent, green sepals. The densely short-hairy achenes have a lobed body projecting into a straight, daggerlike beak. They are borne in a short-cylindrical head.

Ecology & Ethnobotany *Ceratocephala* is Greek for hornhead, and *testiculata* is for two very small sacks at the base of the seeds. Formerly a species in the genus *Ranunculus*.

The species is considered to be invasive and where it is present in large numbers, it is usually an indication of excessive disturbance to the land. In our sagebrush country, it is one of the first plants to flower after the snow melts. This plant occurs on disturbed ground along roads and trails at the lower portions of the Boise Foothills.

CLEMATIS
Clematis hirsutissima & C. ligusticifolia
RANUNCULACEAE

General Description These are herbaceous perennials (*C. hirsutissima*) with erect stems, or woody vines (*C. ligusticifolia*). The leaves are opposite or whorled and simple to pinnately compound. The flowers, lacking petals, are solitary or borne in an open, pyramid-shaped inflorescence. Sepals are petal-like. The species can be found from brushy slopes above creek bottoms to open areas in the upper Boise Foothills.

Comparison of *Clematis* species

C. hirsutissima Plants herbaceous; flowers terminal on main stem
C. ligusticifolia Plants woody, often vines; flowers axillary or rarely on a naked scape from base of plant

Ecology & Ethnobotany The genus is essentially comprised of poisonous species. Many references list *C. ligusticifolia* as poisonous even though the stems and leaves have been chewed by Native Americans as a remedy for colds and sore throats. The fibers in the bark was used for snares and carrying nets. The dried stalks were used in fire-by-friction sets and the feathery seed tails for tinder.

LARKSPUR
Delphinium   RANUNCULACEAE

General Description These are all perennial herbs with tuberous or fibrous roots and erect stems. The leaves are roundish in outline and deeply lobed or divided. The flowers are showy, blue to partly white,

containing 5 petal-like sepals with the uppermost prolonged into a spur. There are also 4 petals, 2 partly enclosed by the upper sepals, the lower 2 often hairy and lobed at the tip. Several species can be found in various habitats of the Boise Foothills.

Ecology & Ethnobotany Cattle and horses can contract the usually fatal disease of delphinosis from eating delphiniums. Plants should be regarded as poisonous.

BRISTLY MOUSETAIL
Myosurus apetalus   RANUNCULACEAE

General Description There are about 15 species of *Myosurus*. They occur in the temperate zones of both the Northern and Southern Hemispheres. This species has stems up to 4 inches tall and leaves up 3 inhes long. The sepal blades are about equal to the spurs and petals are often lacking.

Ecology & Ethnobotany Mousetails are so named for a long, slender column covered with pistils (female seed-bearing organs) that arises from the center of the flower.

Although tiny flies have been observed visiting mousetails, insects apparently are not necessary to transfer pollen. Stone (1959) noted that tiny mousetails are predominantly self- pollinating. Pollen is shed before the flower opens, when the pistils and stamens are covered by the sepals. Fertilization does not take place until 3 to 10 days later, which ensures that pollen will reach all the pistils that have developed. After the pollen is shed, the flower opens. (Stone, D. E. 1959. A unique balanced breeding system in the vernal pool mouse-tails. Evolution 13: 151-174).

BUTTERCUP
Ranunculus   RANUNCULACEAE

General Description Buttercups are either perennials or occasionally annual herbs with simple to compound leaves. The flowers are solitary or borne in a small inflorescence. The 5 petals are normally yellow or white and have a nectar gland at the base. They can be found in many different habitats in the Boise Foothills.

Ecology & Ethnobotany All species are poisonous when raw. The leaves and stems should be boiled in several changes of water to remove the poisonous compounds. The volatile toxin is also rendered harmless by drying. The seeds can be parched and ground into meal for bread or pinole. The roots can

also be boiled and eaten and were an important part of some Native American diets. A yellow dye can be obtained by crushing and washing the flowers. All species have corrosive juice which is very painful if rubbed into the eyes.

WESTERN MEADOW-RUE
Thalictrum occidentale   RANUNCULACEAE

General Description These are rhizomatous herbs with erect stems. The leaves are 2-4 times branched into ultimate leaflets that are shallowly lobed or toothed, and closely resemble the leaves of columbine (*Aquilegia*). There are no petals, and the 4-5 sepals fall soon after opening. They are found in moist areas at various elevations.

Ecology & Ethnobotany The common name meadow-rue is given to *Thalictrum* because of their resemblance to European rues (*Ruta*), which are grown for their aromatic and medicinal properties.

SNOWBRUSH
Ceanothus velutinus   RHAMNACEAE

General Description Snowbrush is easy to identify by its shiny, often sticky, evergreen leaves with 3 main veins. Its small, creamy white flowers are borne in pyramidal clusters. Look for this plant in the upper portions of the Boise Foothills.

Ecology & Ethnobotany The genus has been long recognized as a substitute for commercial black tea and the leaves and flowers could be used to make tea. The seeds can also be used as food. Many species

contain saponin which gives the flowers and fruits their soap-like qualities. The flowers when crushed and rubbed in water, will produce a light lather for purposes of washing oneself. The long, flexible shoots were used in basketry. The red roots yield a red dye.

SERVICEBERRY
Amelanchier alnifolia & A. utahensis   ROSACEAE

General Description These are shrubs or small trees with simple leaves that are serrate on the terminal half. The white flowers have 5 petals and 5 reflexed sepals, and many stamens. The ovary is inferior and the fruit a pome (apple-like). In the Boise Foothills two species may be encountered and all can be used in similar ways.

Comparison of _Amelanchier_ species

Saskatoon serviceberry (_A. alnifolia_) Petals $\frac{1}{2}$ to $\frac{3}{4}$ inch long; styles 5, rarely 4, united below; fruit glabrous; leaves glabrous or almost at maturity

Utah serviceberry (_A. utahensis_) Petals 1/8 to $\frac{1}{4}$ inch long; styles 2-4 in number, rarely 5, and separate to base; fruit hairy; leaves usually hairy, at least underneath at maturity

Ecology & Ethnobotany All species produce edible pomes that ripen in late spring and the summer. They were a considered to be a major food for many Native Americans. In fact, some Native Americans intentionally moved their camps to locations where they

could be more easily harvested. The pomes may be eaten raw, cooked, or dried. After drying, the pomes can be pounded into loaves or cakes. These in turn may be eaten after softening a piece in water or placing them in soups or stews. Prepared this way, the pomes could be kept for several years. Additionally, the dried pomes could be incorporated into pemmican. The wood can be used for arrows, digging sticks, and other useful items.

HAWTHORNE
Crataegus douglasii   ROSACEAE

General Description This is a deciduous shrub or small tree with thorns. The leaves are toothed or lobed and the white flowers are borne in an open inflorescence. The fruits are small pomes, borne in tremendous quantity, and remaining on the tree all winter. *Crataegus* is a large and varied genus containing many species that readily hybridize.

Ecology & Ethnobotany All species produce edible, albeit mealy fruits which may be eaten raw or cooked in small amounts, or dried and mixed into pemmican. A diet high in hawthorne pomes or drinking hawthorne tea is said to reduce weight. The pomes contain a non-toxic heart stimulant and should not be eaten in large amounts or without admixture. The pomes also contain Vitamin C. The thorns have many practical uses such as prongs or rakes, lances for blisters, piercing ears, and as fish hooks.

WILD STRAWBERRY
Fragaria   ROSACEAE

General Description Two species may be found in the upper forested areas of the Boise Foothills, with *F. vesca* more commonly encountered. These white-flowered perennial herbs are produced from rootstocks and have long runners that root at the nodes. The leaves are clustered at the base of the stem and are divided into three egg-shaped, coarsely toothed leaflets. They can be found in moist, humus-rich, well-drained soils of open forest and forest margins.

Woodland strawberry (*F. vesca*) Apical tooth of leaflets greater than those on either side; leaves yellow-green, upper surface bulged between the veins -----

Virginia strawberry (*F. virginiana*) Apical tooth of leaflets smaller than those on either side; leaves blue-green

Ecology & Ethnobotany Strawberries do not keep well and should be dried for future use if not eaten soon after being picked.

Tea made from the green or dried leaves is said to tone up one's appetite. Externally, the leaf tea can also be used as an antiseptic wash for eczema and wounds and as a gargle for sore throat and mouth ulcers. The plants do contain substantial amounts of Vitamins A and C, and sulphur, calcium, potassium, and iron. To remove tartar, rub the berries on your teeth and let the juice sit for a few minutes. Afterwards, brush your teeth thoroughly with baking soda and water.

OCEANSPRAY
Holodiscus discolor   ROSACEAE

General Description This is a deciduous shrub growing up to 10 feet tall with grayish-red bark. The egg-shaped leaves are shallowly lobed with toothed margins. The creamy-white colored flowers are small and borne in a diffusely branched inflorescence. This shrub grows at in the woods or fairly moist areas in the upper portions of the Boise Foothills.

Ecology & Ethnobotany The plant is found in areas prone to wildfire, and it is often the first green shoot to spring up in an area recovering from a burn.

The small dry berries can be used as food, eaten raw or cooked. The hard wood can be used for digging sticks.

MALLOW NINEBARK
Physocarpus malvaceus ROSACEAE

General Description This is a deciduous shrub with nearly glabrous foliage, erect stems growing up to 6 feet tall and having bark that peels back in lines along its axis. The alternate leaves have short petioles and spade-shaped blades (1-2 inches long) that are 3-lobed and coarsely toothed. Leaves are green above and paler beneath. The numerous small bisexual flowers are stalked and borne in a broadly hemispheric inflorescence. The cup-shaped calyx has 5 reflexed lobes and 5 white spreading petals. There are 20-40 stamens and 2-3 styles. The fruit is an inflated 2- to 3-chambered capsule with 2-4 seeds per chamber. This shrub is comnon on rocky slopes and in dry Douglas-fir and ponderosa pine forests in the upper portions of the Boise Foothills.

Ecology & Ethnobotany Mallow ninebark can form dense thickets, which provide good shelter and cover for a variety of wildlife species from small birds to large mammals. Mallow ninebark leaves are palmately three- to five-lobed and begin to turn color as early as late July, becoming brownish-red by early autumn.

"CINQUEFOIL/POTENTILLA" complex
Argentina, Comarum, Dasiphora, Potentilla
ROSACEAE

General Description Cinquefoils are perennial, biennial, or annual herbs, and one shrub with alternate, mostly compound leaves that have membranous appendages at the base of the petioles. The bisexual flowers are solitary or borne in branched, usually open inflorescences. The 5 separate petals are yellow, white, or, in one case, purple, and the 5 sepals have small bracts between them. There are numerous stamens and styles. The fruit is a cluster of achenes borne on the convex receptacle. The lower portion of the calyx is united to the receptacle. They can be encountered in various habitat types at all elevations. Some species of *Potentilla* have been reclassified into other genera.

Ecology & Ethnobotany Cinquefoil is derived from the Latin *quinque* (five) and *folium* (leaf) for the five-parted leaflet, which, however, is not always five-parted. *Potentilla* is derived from the Latin *potentia* meaning power, for some members of the genus were believed to have potent curative powers.

Potentillas are abundant in many habitats and their bright yellow flowers are a common sight to hikers. But because *Potentilla* hybridize, they are often difficult to identify to exact species.

BITTER CHERRY, CHOKECHERRY
Prunus emarginata & P. virginiana   ROSACEAE

General Description Members of this genus are deciduous shrubs or trees that often spread by sending up shoots from shallow roots. The alternate leaves are simple with finely toothed margins. The flowers have a 5-lobed, hemispheric calyx attached to the ovary near the base. The spreading petals are white to greenish or yellowish and larger than the sepals. Each flower has numerous stamens, 1 style, and a 1-celled ovary. The fleshy fruit (cherry-like) contains a single hard seed. Two species can be found in the Boise Foothills.

Comparison of *Prunus* species

Chokecherry (*P. virginiana*) Flowers numerous, in unbranched, long, narrow inflorescences
Bitter cherry (*P. emarginata*) Flowers occur in hemispheric or flat-topped inflorescences

Ecology & Ethnobotany In general, the fruits of all species are sour or bitter when raw, but after cooking the sourness disappears. Native Americans dried the berries whole or in cakes for use in winter. When needed, the dried fruits were soaked in water and then eaten.

To make cakes, the ripe fruits are usually ground up, pits and all, and dried in the sun. When needed, the cakes, or portions thereof, can be soaked in water, mixed with flour and sugar and made into a sauce or gravy. This sauce was eagerly traded among some Native Americans. The only difficulty we've found in preparing cakes in this manner is that the pits do not grind down

nicely into a fine material, leaving larger chunks that could have resulted in broken teeth.

Other uses of the berries included their incorporation into pemmican. They can also be used in making jelly, but because *Prunus* are low in natural pectin, it is advisable to add pectin.

The leaves of the species contain toxic amounts of cyanide as do the seeds (pits). Cyanide is highly volatile and the pits can be rendered safe by long-term drying, by boiling in several changes of water, or by dry roasting. Do not eat them in significant amounts even then unless you mix them with larger quantities of other foods. Prunus shoots, peeled and split, were used in basketry. The wood was used for various implements, such as digging sticks, arrows, and arrow fore shafts.

BITTERBRUSH
Purshia tridentata   ROASCEAE

General Description This fragrant shrub grows up to 8 feet tall. Leaves are deeply 3-cleft into linear lobes, glandular above and hairy below. The leaf margins are rolled inwards. Flowers are pale yellow to white and the fruit is a pubescent oblong achene. Bitterbrush grows on dry slopes and canyons in the Boise Foothills. Flowers from April to June.

Ecology & Ethnobotany The ripe seed coat produces a violet dye. Old *Purshia* stumps produce shredding bark that can be peeled off, worked to soften, and used as toilet paper or material with which to start a fire by friction.

WOOD'S ROSE
Rosa woodsia   ROSACEAE

General Description Woods rose has stems from 20-80 inches tall, with or sometimes without straight or slightly curved prickles at the nodes and scattered prickles between the nodes. The 5-9 elliptical to egg-shaped leaflets and distinctly toothed. Flowers are clustered at the ends of the current season's growth. The glabrous sepals become spreading to erect in fruit, and the deep red to light pink petals. Hips vary from globose to elliptical or pear-shaped. This is the most common rose in our area, occurring as scattered individuals or thickets on open hillsides, along roads, and in river bottoms.

Ecology & Ethnobotany The hips are edible raw, stewed, candied, or made into preserves. They are high in Vitamin C and also contain Vitamins E, B, and K, beta-carotene, calcium, iron, and phosphorus. There are many other edible parts, besides the fruit. Young Rosa shoots in spring make an excellent potherb, and the roots and stems can be used to make a tea. The petals may be used in salads. The peeled spring shoots can also be nibbled upon. Almost all parts of the plant have been made into a wash or dressing for cuts or sores to coagulate blood. One of the more common methods is to sprinkle fine shavings of de-barked stems into a washed wound. The petals can be used as a dressing. A poultice of leaves can be used to relieve insect stings. In addition, the young leaves can be washed, cut into small pieces, and dried for a hot tea.

BLACKBERRY, RASPBERRY, THIMBLEBERRY
Rubus leucodermis & R. parviflorus   ROSACEAE

General Description These species are deciduous shrubs with arching or trailing stems covered with bristles and prickles. The flowers have white petals and the fruit is a coherent cluster of small, 1-seeded drupes (raspberries, blackberries, dewberries, cloudberries, marionberries).

Comparison of *Rubus* species

Thimbleberry (*R. parviflorus*) Leaves lobed but not divided into leaflets; stems unarmed
Whitebark Raspberry (*R. leucodermis*) Leaves divided into leaflets; stems with thorns or prickles

Ecology & Ethnobotany All species produce edible fruits. Flowers can be added to salads and can be nibbled upon when hiking. The fresh or dried leaves can be steeped for a tea, alone or in herbal blends. Do not use the wilted or molded foliage, as it may be toxic. The young shoots cut just above ground can be peeled and eaten raw or cooked.

MOUNTAIN ASH
Sorbus scopulina   ROSACEAE

General Description This species is a shrub up to 13 feet tall with reddish-brown branches and buds and young twigs that are finely white- or grayish-hairy. The leaves have 11-17 shiny and nearly glabrous, broadly lance-shaped leaflets with finely toothed margins. The narrow appendages at the base of the petioles fall early in the season. The white, oval petals are about $\frac{1}{4}$ inch long, and the berries are orange to red and shiny but without a waxy coating. It can be found in moist meadows and forest openings in the upper portions of the Boise Foothills.

Ecology & Ethnobotany The fruits may be eaten raw, cooked, or dried. They are high in Vitamin A and C, and carbohydrates. Unripe berries are very bitter and somewhat unpalatable. The fruits, which are pomes, are commonly processed into jams and jellies.

They have high pectin content and jell readily. As a coffee substitute, grind the dried, roasted seeds. The berry juice can be used as a gargle for sore throat and as an antiseptic wash for cuts. sorbitol, the sugar in the fruit of *Sorbus* is being used commercially for sweetening candies, toothpaste, and other products.

SPIRAEA, MEADOWSWEET
Spiraea lucida   ROSACEAE

General Description This spiraea has erect or spreading stems growing up to 2 feet tall, arising from strong rhizomes. The oblong to egg-shaped leaves (1-3 inches long) are dark green above and pale beneath. The flat- topped inflorescence bears flowers that are cream to white, sometimes with a faint pinkish tinge. The petals are about 1/8-inch long, and the fruits are about 1/8-inch high.

Ecology & Ethnobotany *Spiraea* is a source of methyl salicylate, similar to the active ingredient in aspirin. Native Americans brewed a tea from the stem, leaves, and flowers of some species to use as a pain reliever. The plants are astringent and a poultice made from the leaves and bark was used to treat ulcers, burns, and tumors. The roots were also peeled and boiled until soft, mashed and used as a poultice for burns. The wiry, branching twigs can be used to make broom-like implements for collecting tubers.

BEDSTRAW, CLEAVERS
Galium aparine   RUBIACEAE

General Description Despite their small flowers, the various species of *Galium* are unmistakable. They are annual or perennial herbs with 4-angled stems and whorled leaves. The small, 4-parted flowers are white or greenish and the fruits are smooth or bristly hairy. They can be found in various habitats from the low to higher elevations.

Ecology & Ethnobotany None of the species of *Galium* are known to be poisonous. Although *G. aparine* is the most commonly used species, it is believed that all other species can be used similarly. The very young leaves and stems can be used as a potherb. The small hairs on the

stems make the plant difficult to swallow raw, boiling or steaming, however, does soften them up. If the stems are too fibrous, use only the leaves. Slow roasted until dark brown and ground, the ripe fruit can be used as a coffee substitute.

ASPEN, COTTONWOOD
Populus   SALICACEAE

General Description Cottonwoods are trees with sticky, resinous leaf buds, and deciduous leaves. Older trees of some species have gray, rough bark; young bark is smooth and whitish. The flowers are borne in catkins that appear before the leaves. Cottonwoods are usually associated with streams.

1a. Bark smooth; petioles strongly laterally flattened below blade ----- **aspen (*P. tremuloides*)**
1b. Bark rough; petioles not flattened ----- **2a**

2a. Petioles less than 1/3 the blade length; blades lanceolate ----- **narrowleaf cottonwood (*P. angustifolia*)**
2b. Petioles over 1/3 the blade length; leaves ovate ----- **black cottonwood (*P. balsamifera* ssp. *trichocarpa*)**

Ecology & Ethnobotany The catkins may be eaten raw or boiled in stews and are a source of Vitamin C. The inner bark can also be eaten as a spring tonic, or dried and ground into a flour substitute or extender. The fresh or dried plant can be used in poultices for muscle aches, sprains, or swollen joints. The primary action of *Populus* is that of an analgesic, used topically and internally. It contains varying amounts of populin and salicin, compounds related to early forms of aspirin. The leaves and bark are most effective parts for tea and aid in diarrhea problems. The wood makes for an excellent bow and drill fire set. Cottonwoods are considered to be botanical indicators of water.

WILLOWS
Salix   SALICACEAE

General Description Many species of willow can be found from along the Boise River upwards into the foothills and on the forest. They are mostly shrubs with numerous stems. Flowers are in catkins that appear before, with, or after the leaves. Willows generally grow along streams or other moist habitats.

Ecology & Ethnobotany The young shoots and leaves can be eaten raw. The bitter inner bark can also be eaten raw, although it is better dried and ground into flour substitute or extender. The plant contains salicin which is similar to aspirin and useful as a substitute. Any part of the willow can be used to produce a tea for use as an aspirin replacement for headache and body pain. The highest concentrations of salicin, however, are found in the inner bark. Because it is not nearly as strong as aspirin, you may have to drink quite a bit of it.

The leaves have astringent properties that are effective when placed on wounds and cuts. Bark was chewed as a toothache remedy. Bark, leaves, twigs, and roots produced medicinal teas, powders, washes, and poultices to relieve pain, swelling, infection, bleeding, and many other ailments. Willows, like the cottonwoods, are botanical indicators of water. The branches of many willow species are very flexible and make them very useful for traps, arrow shafts, and other needs, such as basketry. Fiber from bark was used for cordage, nets, and clothing.

BASTARD TOADFLAX
Comandra umbellata   SANTALACEAE

General Description Bastard toadflax is a partially parasitic perennial herb with a waxy surface and a rather woody base. The leaves are linear and the flowers are bell-shaped. The fruit is a 1-seeded, berry-like drupe. It is common and widespread in the forest above the Boise Foothills. The roots are blue when cut.

Ecology & Ethnobotany The mature, brown, urn-shaped fruit of bastard toadflax may be eaten raw and is best when slightly green. They were popular with Native Americans because of their sweet taste. The berries, however, are rarely found in sufficient quantities for more than a pleasant tidbit. Consuming too many berries may cause nausea.

WOODLANDSTAR
Lithophragma   SAXIFRAGACEAE

General Description Three species may be found in the Boise Foothills and include bulbous woodlandstar (*L. glabrum*), smallflower woodlandstar (*L. parviflorum*) slender woodlandstar (*L. tenellum*)

In short, these are slender, perennial herbs with usually glandular, unbranched stems. The rootstocks often bear bulbs. The mostly basal leaves have petioles with a swollen base and divided or lobed blades. The showy, short-stalked flowers are borne in a flat-topped to narrow, few-flowered inflorescence. The calyx is cup- or bell-shaped with 5 shallow lobes. The 5 white to pink petals are lobed or deeply divided. There are 10 stamens and 3 styles. The fruit is a 3-chambered capsule. They are found in moist or shaded places.

Ecology & Ethnobotany There are no recorded uses for Idaho species, but the roots of *L. affinis* were chewed by the Native Americans in California to treat stomach ailments and colds.

ALUMROOT
Heuchera cylindrica & H. grossularifolia
SAXIFRAGACEAE

General Description Members of this genus are perennial herbs with slender, erect, leaf less stems from scaly, creeping rootstocks and a branched crown. The basal beaves have petioles with membranous appendages fused to the base and blades that are palmately lobed and toothed. The short-stalked flowers are borne in long, narrow, terminal inflorescences. The saucer- to bell-shaped calyx has 5 lobes, and the 5 small petals (in some species may be lacking) are entire-margined and greenish, white, or pinkish. The 5 stamens are opposite the sepals. Fruit is a capsule with 2 elongated, divergent, hollow beaks. These plants occur in various habitats that include moist soils and rocky areas up to the alpine zone.

Comparison of *Heuchera* species

Gooseberry alumroot (*H. grossulariifolia*) Petals 5, present on all flowers and nearly as long or longer than the sepals

Roundleaf alumroot (*H. cylindrica*) Petals often less 5 or lacking, usually ½ as long as sepal

Ecology & Ethnobotany The leaves of all species are edible, although they are not choice. They have a sour taste because of the high tannin content. Therefore, the leaves should be boiled or steamed. Since they are rather tough, we found them to be more palatable if chopped and added to soups or salads.

Heuchera is said to be one of the strongest astringents due to their high tannin content, as much as 20 percent their weight in tannins. Tannins tend to shrink swollen, moist tissues. As such, alumroots are also gastrointestinal irritants and have been known to cause kidney and liver failure. Ingestion of the plant should be in moderation. Otherwise, the pounded, dried roots of many species have been used as a poultice that stops bleeding and promotes healing when applied to cuts and sores. The raw root, eaten in small amounts, has been used as a cure for diarrhea. A tea from the roots can also be used as a gargle for sore throats. The powdered roots have been used as an antiseptic.

SAXIFRAGE
Saxifraga   SAXIFRAGACEAE

General Description Two species in the Boise Foothills – peak saxifrage and diamondleaf saxifrage. In recent years, research has reclassified many species of the genus *Saxifraga* into *Micranthes*. In general, they are perennial herbs with basal, alternate, or opposite leaves. The flowers are broadly bell-shaped.

Ecology & Ethnobotany The generic name is from Latin *saxum* meaning rock and *frangere*, meaning to break, and alludes to many species rocky habitat.

The genera as a whole are regarded as a safe group of plants. The leaves can be used fresh or in stews and are high in Vitamins A and C. In China, some species were used in the treatment of nausea and ear infections. In our area, there is little documentation regarding medicinal uses.

PAINTBRUSH
Castilleja   SCROPHULARIACEAE

General Description This is a large genus found primarily in western North America that contains many species, several of which occur in the Boise Foothills. The genus is easily recognized, but many species are notoriously difficult to identify. They are perennials with deeply lobed to entire leaves. The flowers are subtended by colorful leaf-like bracts. Some paintbrushes are partial root parasites found in various habitats up to the alpine zone.

Ecology & Ethnobotany Many, if not all of the species have flowers and bracts that can be eaten raw. The seeds of some species were gathered, winnowed, dried, and stored for winter use. In winter they were parched, pounded and eaten dry. The plants, however, absorb selenium from the soil and so should be taken in moderation. Symptoms in humans of selenium poisoning will vary with the amount and form ingested, but may include difficulty in breathing, excessive urine production, loss of appetite, mental depression, a weak and rapid pulse, blurry vision, digestive upset, and eventually coma and death.

BLUE-EYED MARY
Collinsia parviflora   SCROPHULARIACEAE

General Description Blue-eyed Mary is a small annual with mostly branched stems, generally less than 5 inches tall. The entire-margined, broadly lance-shaped leaves are opposite each other on the stem. Herbage is most often sparsely hairy with glands in the inflorescence. Small flowers are borne on long, weak stalks in the axils of the upper Leaves and Leaflike bracts. The blue, tubular corolla with a white, 2-lobed, upper lip and a 3-lobed lower lip. There are 4 stamens, and the fruit is an ellipsoid, many-seeded capsule. This species is common in moist habitats of the Boise Foothills.

Ecology & Ethnobotany The genus name honors Zaccheus Collins, a Philadelphia botanist that lived from 1764 to 1831. Thomas Nuttall named this genus and David Douglas (of Douglas-fir fame) named the species in 1827.

YELLOW MONKEYFLOWER
Mimulus guttatus   SCROPHULARIACEAE

General Description Several species of monkeyflower can be found in the Boise Foothills. They are annual or perennial herbs with opposite leaves. The flowers flare at the mouth to form 5 lobes, 2 which form the upper lip and 3 lobes that form the lower.

Yellow monkeyflower is a perennial growing up to 40 inches tall and has hollow stems. Leaves are opposite, oval to rounded in shape, finely toothed, and 3/8 to 3 inches long. The lower leaves are long-petioled, while the upper leaves are sessile and slightly fused together at their bases. Flowers are on pedicels and are $\frac{3}{4}$ to $2\frac{1}{2}$ inches long. Calyx is bell-shaped, inflated, and is tinged with red and 5-angled, 5/8 to 1-inch long. The yellow corolla is 2-lipped, with red spots, 5/8 to $1\frac{1}{2}$ inches long. Yellow monkeyflower is common along streams and in wet places and can be found

throughout the area. Flowers from March to August.

Ecology & Ethnobotany The young stems and leaves of *M. guttatus* have been used as salad greens. Sometimes, leaves were burned and the ash used a salt. Weedon (1996) indicates that the young herbage of *Mimulus* species may be eaten in salads, and that they grow bitter with age, but remain edible.

PENSTEMON
Penstemon   SCROPHULARIACEAE

General Description There are many species of Penstemon in the area. In general, they are perennial herbs with opposite leaves. The flower is strongly to indistinctly 2-lipped at the mouth with a 2-lobed upper lip and a lower lip with three lobes. There are four anther-bearing (fertile) stamens and a single sterile stamen that is often hairy at the tip. The fruit is a many seeded capsule. Penstemons occur in dry or moist meadows or forest openings up into the alpine zone.

Ecology & Ethnobotany The genus name is from the Greek *pete*, meaning five, and *stemon*, meaning thread, referring to the slender fifth stamen.

Genetic research in the 1990s and early 2000s showed that the *Penstemon* genus belongs in Plantaginaceae (Plantain Family), not Scrophulariaceae. In

North America there are over 250 *Penstemon* species, which ties it with *Eriogonum* (buckwheat) for third most numerous. *Carex* (sedge) has 480 and *Astragalus* (milkvetch) 350 species.

TAILED KITTENTAILS
Synthyris missurica   SCROPHULARIACEAE

General Description Kittentails are native perennial herbs with short rhizomes, petiolate basal leaves, and small, alternate, leaf like bracts on the stem. The short-stalked flowers, often subtended by a small bract, are borne in a congested, spike-like inflorescence at the ends of the stems. The 4-lobed corollas are bowl-shaped, and the upper lobe is larger than the others. There are 2 stamens that protrude well beyond the mouth of the corolla. Look for this species in the upper portion of the Boise Foothills.

Ecology & Ethnobotany The common name of "kittentails" for this showy, deep blue wildflower comes from the elongated raceme that, with ample imagination, could resemble the tail of a kitten. The two stamens that protrude beyond the petals give the entire inflorescence a furry ambiance. The genus name is from the Greek *syn* (with) and *thyris* (window), alluding to the valves of the capsule. Several species are grown as ornamentals.

WOOLLY MULLEIN
Verbascum blattaria & V. thapsus
SCROPHULARIACEAE

General Description These are introduced, Eurasian biennial herbs that produce rosettes of basal leaves the first year and solitary, flowering stems with alternate leaves in the second or third year. The stalked flowers are borne in an open or congested, spikelike, terminal inflorescence. The yellow or white, saucer-

shaped corollas are 5-lobed with the upper 2 lobes slightly shorter than the others. There are 5 stamens. Both are rather common species in the Boise Foothills.

Ecology & Ethnobotany The leaves of mullein (*V. thapsus*) are said to be edible when eaten in small quantities and cooked. Because of their woolly texture, however, we have found the plants to be undesirable. The leaves can also be used as poultices applied locally to hemorrhoids, sunburn, and inflammations. The dried stalks are ideal for use as hand-drills to start fires. The flowers and leaves produce yellow dye; as a toilet paper substitute, the large fresh leaves are choice.

SPEEDWELL

Veronica   SCROPHULARIACEAE

General Description Several species of speedwell can be found in the Boise Foothills. They are annual or perennial herbs with opposite, alternate, or, rarely, whorled leaves. The small, blue, pink, lilac, or white flowers are saucer-shaped and 4-lobed, the upper lobe being the largest. The mature fruit is often necessary for identification. Most species are found in wet soils or shallow waters from low to high elevations.

Ecology & Ethnobotany The leaves and stems of all species, when collected during the spring and early summer, can be eaten like watercress, added to salads, or prepared as potherbs. The taste of the various species ranges from spicy to bitter to bland depending on personal taste. The plants also contain moderate amounts of Vitamin C and were once used to prevent scurvy. The leaves and stems can also be steeped as a tea. Care should be taken to avoid plants growing in polluted waters.

NIGHTSHADE

Solanum   SOLANACEAE

General Description This is a highly diverse genus comprising more than a thousand species worldwide. In general, nightshades are annual or perennial herbs with flowers that resemble those of tomato or potato plants.

The fruit is a many-seeded berry, surrounded in part by the persistent calyx. The various nightshades can be found from moist, open habitats to disturbed, waste areas at the lower elevations in the Boise Foothills.

Ecology & Ethnobotany *Solanum* probably comes from the Latin *solamen*, which means quieting, referring to the sedative properties of some species. Various species of *Solanum* use "buzz pollination" (sonication) – when various large bumblebees (*Bombus*) and carpenter bees (*Xylocopa*) come a visiting. Interesting factoid, only approximately 8% of the flowers in the world are primarily pollinated by this type of pollination system.

BROADLEAF CATTAIL
Typha latifolia   TYPHACEAE

General Description Cattails are found over much of North America. A cattail plant produces a basal cluster of narrow, ribbon-like leaves that are several feet long and stand almost vertically. The upright stem is unbranched, not quite as long as the leaves, and it bears a long, dense, brown spike at the upper end. The spike may vary from 4 to more than 12 inches long. Its upper part bears stamens intermixed with long hairs, each stamen constituting a flower, while its lower part bears pistillate flowers, each flower consisting only of an ovary with an abundance of dark hairs at its base. A second species known as narrowleaf cattail (*T. angustifolia*) may also be encountered. It can be used similarly. The following key may be useful in identifying the species.

Comparison of *Typha* species

T. latifolia Staminate and pistillate parts of spike usually contiguous or almost; stigmas oblanceolate to obovate in shape; leaves over $\frac{1}{2}$ inch wide

T. angustifolium Staminate and pistillate parts of spike separated by at least $\frac{1}{4}$ inch; stigmas linear; leaves less than $\frac{1}{2}$ inch wide

Ecology & Ethnobotany Female cattail flowers mature before male flowers, making cross pollination possible or even likely.

Virtually every part of these plants has a use, from food to fiber. In fact, Native Americans and wilderness adventurers consider the cattail the "supermarket of the swamps." Although both cattail species have edible rhizomes, the rhizomes should never be raw since they may cause vomiting. The rhizomes should be boiled or roasted or dried and then ground into meal or flour.

When pulling up the rhizome, you may notice newly emerging buds. These can be scrubbed, peeled, and eaten raw or boiled. The swollen joint between bud and rhizome is also starchy. Peel it, then roast or boil for a potato-like vegetable. Like the rhizomes, this part should not be eaten raw. The young green shoots can be peeled of their green outer layer and eaten raw or cooked. It is always good to boil them in a couple of changes of water if there is any bitterness. The peeled core can also be sliced and added to salads.

Useful fibers can be derived from cattails. Fibers in stems can be loosened by soaking plant material in water for several days. The silky fluff on the seeds is buoyant and water repellent and makes a good insulator, especially in boots. The silk can be used for stuffing items from pillows to down vests. It can also be used for tinder. The fuzz will explode into flame with a spark from a flint and steel set. Leaves can be woven to make mats, sandals, baskets, etc. The stems provide a good coil foundation for baskets. Additionally, the stalks have been used as arrows and hand drills. A toothbrush can be fashioned from the fuzzy stem with the flowers removed.

NETLEAF HACKBERRY
Celtis laevigata var. *reticulata*   ULMACEAE

General Description This is a small tree or shrub with leaves that are ovate to lanceolate in shape, with entire to serrate edges. The fruit is a drupe, and the plants can usually be found growing along streams or on dry canyon slopes at the lower elevations within the Boise Foothills.

Ecology & Ethnobotany The small orange, red, or yellow fruits are edible raw and have a sweet taste to them. The entire fruits can also be dried and then ground into a flour.

STINGING NETTLE
Urtica dioica   URTICACEAE

General Description This is an annual or perennial herb with stinging hairs. The flowers are numerous, small, and clustered on drooping branches at the base of the leaves. Within the Boise Foothills, stinging nettles can be found along roadsides, streams, in moist areas and waste places in the low to middle elevations. Stinging nettle is an indicator of good soil conditions.

Ecology & Ethnobotany The young stems and leaves of stinging nettle are edible after boiling and are very delicious as a spinach substitute. Boiling the leaves destroys the formic acid found in the hairs. The leaves are high in vitamins A, C, and D, the latter of which is rare in plants. The roots are also edible after they have been roasted.

The older stems become fibrous, which reduces their edible qualities, but allows them to be used to produce strong cordage. The older leaves also contain cystoliths that can irritate the kidneys. A yellow dye may be obtained by boiling the roots.

LONGHORN PLECTRITIS
Plectritis macrocera   VALERIANACEAE

General Description This is a small, early blooming annual with slender to stout stems from 2 to 6 inches tall. The herbage is generally smooth, although it maybe finely glandular in the inflorescence. The stems are single with

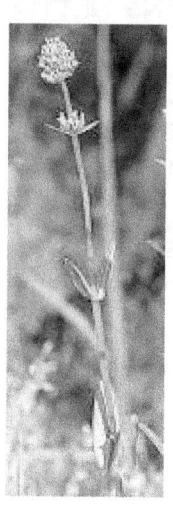

one to several pair of clasping leaves rising from a basal rosette of tiny, oval leaves. The basal leaves often turn yellow very quickly. They are obovate and short petiolate while the stem leaves are oblong or elliptic and sessile. The leaves range from ½ to 1¾ inch long and 1/8 to ¾ inch wide. The inflorescence is clusters of cylindrical flowers at the apex of the stem. The flowers are 5-petaled and white or pinkish. The corolla is about ¼ inch long with a short, thick spur.

Ecology & Ethnobotany Look for this plant in moist, open or partially shaded habitats in the lower portions of the Boise Foothills.

VERBENA
Verbena bracteata & V. hastata   VERBANACEAE

General Description Members of this genus are perennials with opposite, toothed leaves. Flowers, each subtended by a narrow bract, are borne in terminal, sometimes branched spikes. The calyx is 5- lobed, and the corolla is tubular with 5 flaring lobes. There are usually 4 stamens and a single style. The fruit is a cluster of 4 nutlets.

Comparison of _Verbena_ species

V. bracteata Flowers greatly surpassed by subtending bracts; stems lax or prostrate
V. hastata Flowers longer than subtending bracts

Ecology & Ethnobotany The seeds of *V. hastata* may be gathered, roasted, and ground into bitter tasting flour. Leaching the flour may remove the bitter taste.

VIOLET
Viola   VIOLACEAE

General Description Violets are low-growing, perennial or annual herbs. The leaves are spade-shaped and basal. The flowers occur singly on the ends of stems and have five petals. There are 2 upper and 2 lateral petals, and 1 lower petal that is prolonged into a nectar holding pouch at the base of the flower. Most species also have small, self-fertilizing flowers that do not open. In the Boise Foothills violets can be found in meadows and open forests.

Ecology & Ethnobotany The leaves, buds, and flowers of possibly all species are edible raw or cooked, with some being more palatable than others; the leaves make a good tea. Adding the leaves to soups make them thicker. Violets are high in Vitamin C and beta-carotene.

Collect the plants by leaving the roots intact. Since many species reproduce vegetatively, you will probably not inhibit next year's growth significantly. Many naturalists indicate that all violets are safe for consumption, but there are some experts that insist some yellow species may be somewhat purgative. All species do have a tendency to be slightly laxative, so proceed slowly. The flowers have also been candied or made into jellies and jams.

PUNTUREVINE, GOATHEAD
Tribulus terrestris   ZYGOPHYLLACEAE

General Description Puncture vine is an introduced annual plant with trailing, hairy stems, and an extensive root system. The leaves are opposite and pinnate with four to eight pairs of leaflets. The flowers are yellow and are borne singly in the leaf axils. The fruit is hard, consisting of five spiny nutlets or burrs that break apart into five "tack-like" sections upon maturity. These burs may injure livestock and are the bane of

bicyclists. The plant is found in disturbed and waste places at the lower elevations in the Boise Foothills

Ecology & Ethnobotany Puncture vine has a 5,000-year history of medicinal uses, particularly in China and India. It was used for boosting the hormone production in men and women, and for urinary tract problems, itchy skin, and blood purification. The stems of the plant are considered to be astringent and act upon the mucous membrane of the urinary tract. A tea from the aboveground part of the plant is said to be good for arthritis.

We don't stop hiking because we grow old - we grow old because we stop hiking.

-- Finis Mitchell

INDEX

A

Abronia · 151
Acer · 15
Achillea · 31
Achnatherum · 169
Aconitum · 125
Actaea · 184
ADOXACEAE · 94
Aegopodium · 19
Agastache · 131
Agoseris · 31
Agropyron · 163
Agrostis · 164
ALDER · 70
ALFALFA · 119
ALLIACEAE · 136
Allium · 136, 145
Alnus · 70
ALUMROOT · 207
Alyssum · 79
ALYSSUM · 79
AMARANTH · 16
Amaranthus · 16
Ambrosia · 32, 33, 104
Amelanchier · 191
Amsinckia · 72
ANACARDIACEAE · 17, 18

Anaphalis · 33
Antennaria · 34
Anthemis · 34, 59
APIACEAE · 20, 22, 23, 25, 26, 27
APOCYNACEAE · 28, 29
Apocynum · 28
Aquilegia · 185
Arabis · 79
Arctium · 35
Arenaria · 95, 96, 98
Argentina · 196
Arnica · 36
ARNICA · 36
Artemisia · 37, 38, 158
Asclepias · 29
ASH · 201
ASPEN · 204
Asperugo · 73
ASTER · 51, 57
ASTERACEAE · 31, 32, 33, 34, 35, 36, 37, 38, 39, 41, 42, 43, 44, 45, 47, 48, 49, 50, 51, 52, 53, 54, 55, 56, 57, 58, 59, 60, 61, 62, 63, 64, 65, 66, 67, 68
Astragalus · 116, 213
Atriplex · 101

Avena · 165

B

Balsamorhiza · 39
BALSAMROOT · 39
BANEBERRY · 184
Barbarea · 80
BARBERRY · 69
BARLEY · 168
Bassia · 102
BEDSTRAW · 203
BEEPLANT · 92
BENTGRASS · 164
BERBERIDACEAE · 69
Berteroa · 81
Betula · 71
BETULACEAE · 70, 71
BINDWEED · 108
BIRCH · 71
BISTORT · 175
Bistorta · 175, 177
BITTERBRUSH · 198
BITTERCRESS · 84
BITTERROOT · 180
BLACKBERRY · 200
BLADDERFERN · 113
BLAZINGSTAR · 148
BLUE FLAG · 129
BLUE GRASS · 168
BLUEBELLS · 76
BLUE-EYED MARY · 210

Boechera · 79
BORAGINACEAE · 72, 73, 74, 75, 76, 77, 78, 126, 128
BOUNCING BET · 99
Brassica · 82
BRASSICACEAE · 79, 80, 81, 82, 83, 84, 85, 86, 87, 88, 89, 90, 91, 92
Brodiaea · 145
BROME GRASS · 165
Bromus · 165
BROOM-RAPE · 157
BUCKWHEAT · 175
BUGLEWEED · 133
Buglossoides · 75
BUR BUTTERCUP · 186
BURDOCK · 35
BUTTERCUP · 189

C

Calandrinia · 179, 180
Calochortus · 138, 145
CAMAS · 144
Camassia · 144
Camelina · 83
Camissonia · 152
CAPRIFOLIACEAE · 93, 94
Capsella · 83
Cardamine · 84

Cardaria · 85
Carduus · 41
Carex · 112
CARROT · 23
CARYOPHYLLACEAE ·
   95, 96, 97, 98, 99,
   100
Castilleja · 209
CATCHFLY · 99
CATNIP · 134
CATTAIL · 216
Ceanothus · 190
Celtis · 218
Centaurea · 2, 41
Cerastium · 96
Ceratocephala · 186
Chaenactis · 42
Chamaesyce · 115
Chamerion · 152
chamomile · 59
CHAMOMILE · 34
CHEESEWEED · 149
CHENOPODIACEAE ·
   101, 102, 104, 105,
   106, 107
Chenopodium · 102
CHERRY · 197
CHICKWEED · 97, 100
CHICORY · 45
CHOKECHERRY · 197
Chondrilla · 43
Chorispora · 85
Chrysothamnus · 44, 49
Cichorium · 45

Cicuta · 20, 21, 25
Cinquefoil · 196
Circaea · 154
Cirsium · 41, 45, 106
Clarkia · 152, 154
CLARKIA · 154
Claytonia · 181, 182
CLEAVERS · 203
Clematis · 186
CLEMATIS · 186
Cleome · 92
CLIFF FERN · 113
CLOVER · 120
COCKLEBUR · 68
Collinsia · 210
Collomia · 170
COLLOMIA · 170
COLUMBINE · 185
Comandra · 206
COMANDRACEAE ·
   206
Comarum · 196
Combseed · 78
Common Kochia · 102
CONEFLOWER · 62
Conium · 22
CONVOLVULACEAE ·
   108
Convolvulus · 108
Conyza · 47
Coreopsis · 47
CORNACEAE · 109
Cornus · 109

COTTONBATTING PLANT · 62
COTTONWOOD · 204
COWPARSNIP · 23
CRASSULACEAE · 110
Crataegus · 193
Crepis · 48
Crossflower · 85, 86
Croton · 115
Cryptantha · 73, 74, 78
CRYPTANTHA · 73
CUDWEED · 51
CURRANT · 125
Cuscuta · 111
CUSCUTCEAE · 111
CYPERACEAE · 112
Cystopteris · 113

D

DANDELION · 65
Dasiphora · 196
Daucus · 23
Delphinium · 188
Dentaria · 84
Deschampsia · 166
Descurainia · 86
Dipsacus · 112
Disporum · 142
DISPSACACEAE · 112
DOCK · 178
DODDER · 111
DOGBANE · 28

DOGWOOD · 109
Draba · 87
DRABA · 87
DROPS OF GOLD · 142
DRYOPTERIDACEAE · 113
DUSTYMAIDEN · 42
Dysphania · 104

E

ELDERBERRY · 94
Elymus · 167
ENCHANTER'S NIGHTSHADE · 154
Epilobium · 152, 155
EQUISETACEAE · 114
Equisetum · 114
Eremocarpus · 115
ERICACEAE · 150
Ericameria · 49
Erigeron · 47, 49
Eriogonum · 158, 175, 176
Eriophyllum · 50
Erodium · 123, 124
Erysimum · 87
Erythronium · 139
Eucephalus · 51
Euphorbia · 115
EUPHORBIACEAE · 115

EVENING-PRIMROSE · 152, 156
EVERLASTING · 34

F

Fabaceae · 117
FABACEAE · 116, 118, 119, 120, 121
FALSE DANDELION · 60
False Flax · 83
FESCUE GRASSES · 167
Festuca · 167
FIDDLENECK · 72
FIR · 161
FIREWEED · 152
FLAX · 147
FLEABANE · 49
FORGET-ME-NOT · 77
Fragaria · 193
Frasera · 122
FRINGE-POD · 92
Fritillaria · 140, 145
FRITILLARY · 140

G

Galium · 203
Gayophytum · 155, 156

GENTIANACEAE · 122
GERANIACEAE · 123, 124
Geranium · 124, 125
GERANIUM · 124
GILIA · 171
GLACIER LILY · 139
Glechoma · 131
GLOBEMALLOW · 150
Glycyrrhiza · 117
Gnaphalium · 51, 62
GOAT'S BEARD · 66
GOATHEAD · 222
GOLDENBUSH · 49
GOLDENROD · 63
GOOSEBERRY · 125
GOOSEFOOT · 102
GOUTWEED · 19
GRASSWIDOW · 130
Grayia · 104
GREASEWOOD · 107
Grindelia · 52
GROMWELL · 75
GROSSULARIACEAE · 125
GROUND IVY · 131
GROUNDSEL · 61
GROUNDSMOKE · 155
Gumweed · 52
Gutierrezia · 52
Gymnosteris · 171
GYMNOSTERIS · 171

H

HACKBERRY · 218
Hackelia · 74
HAIRGRASS · 166
Haplopappus · 49
HAWKSBEARD · 48
HAWKWEED · 54
HAWTHORNE · 193
Helianthus · 53, 68
HENBIT · 132
Heracleum · 23
Heuchera · 207, 208
Hieracium · 54
HOLLYHOCK · 148
Holodiscus · 194
Holosteum · 97
HONEYSUCKLE · 93
HOPSAGE · 104
Hordeum · 168
HOREHOUND · 133
HORSE MINT · 131
HORSETAIL · 114
HORSEWEED · 47
HYDRANGEACEAE · 126
HYDROPHYLLACEAE · 128
Hydrophyllum · 126
HYPERICACEAE · 107

I

Iliamna · 148
Indian hemp · 29
INDIAN POTATO · 26
INDIAN RICE GRASS · 169
Ipomopsis · 171
IRIDACEAE · 129, 130
Iris · 129
Iva · 55

J

JERUSALEM OAK GOOSEFOOT · 104

K

KITTENTAILS · 213
KNAPWEED · 41
KNOTWEED · 177
Kochia · 102
Krascheninnikovia · 105

L

Lactuca · 56
LAMIACEAE · 131, 132, 133, 134, 135, 136

Lamium · 132
LARKSPUR · 188
Lathyrus · 121
Layia · 56, 57
LAYIA · 56
Lepidium · 88
Leptodactylon · 173
LETTUCE · 56
Lewisia · 180
LICORICE · 117
LILIACEAE · 138, 139,
    140, 142, 143
LILY · 141, 146
LINACEAE · 147
Linanthus · 173
Linum · 147
Lithophragma · 206
Lithospermum · 75, 76
LOASACEAE · 148
Lomatium · 25
LOMATIUM · 25
Lonicera · 93
LUNGWORT · 76
LUPINE · 118
Lupinus · 118
Lychnis · 97
Lycopus · 133

M

Machaeranthera · 57
Madia · 58
MADWORT · 73, 79, 81

Mahonia · 69, 70
Maianthemum · 141
MALLOW · 149
Malva · 149
MALVACEAE · 148,
    149, 150
MAPLE · 15
MARIPOSA LILY · 138
Marrubium · 133
Matricaria · 58
MEADOW-RUE · 190
MEADOWSWEET ·
    202
Medicago · 119
MELANTHIACEAE ·
    144, 146
Melilotus · 120
Mentha · 134
Mentzelia · 148
Mertensia · 76
Micranthes · 209
Microseris · 59
Microsteris · 173
MILKVETCH · 116
MILKWEED · 29
Mimulus · 211, 212
MINT · 134
MOCKORANGE · 126
Moehringia · 98, 99
MONKEYFLOWER · 211
Monolepis · 105
Montia · 181, 182, 183
MONTIACEAE · 179,
    180, 181

Montiastrum · 181
MOUNTAIN DANDELION · 31
MOUSE EARS · 96
MOUSETAIL · 188
MUGWORT · 38
MULLEIN · 214
MUSTARD · 82
Myostis · 77
Myosurus · 188

N

Nasturtium · 89
Nemophila · 128
NEMOPHILA · 128
Nepeta · 134
NETTLE · 219
NIGHTSHADE · 215
NINEBARK · 195
Noccaea · 90
Nodding Silverpuff · 59
Nothocalais · 60
NYCTAGINACEAE · 151

O

OATS · 165
OCEANSPRAY · 194
Oenothera · 152, 156

Olsynium · 130
ONAGRACEAE · 152, 154, 155, 156
ONION · 136
Onopordum · 60
ORCHIDACEAE · 157
OROBANCHACEAE · 157, 209
Orobanche · 157
Orogenia · 26
Oryzopsis · 169
Osmorhiza · 27, 185

P

Packera · 61
Paeonia · 158
PAEONIACEAE · 158
PAINTBRUSH · 209
PARSLEY · 25
Pectocarya · 78
PENNYCRESS · 90
Penstemon · 212
PENSTEMON · 212
PEONY · 158
PEPPERGRASS · 88
PEPPERWEED · 88
Perideridia · 27
Persicaria · 177
Phacelia · 128
PHACELIA · 128
Philadelphus · 126
Phlox · 173, 174

PHLOX · 173, 174
PHRYMACEAE · 211
Physocarpus · 195
PIGWEED · 16
PINACEAE · 159, 161
PINE · 159
PINEAPPLE WEED · 58
PINEDROPS · 150
Pinus · 150, 159
Piperia · 157
Plagiobothrys · 78
PLANTAGINACEAE · 163, 215
Plantago · 163
PLANTAIN · 163
Plectritis · 220
PLECTRITIS · 220
Plumeless Thistle · 41
Poa · 168
POACEAE · 163, 164, 165, 166, 167, 168, 169
POISON HEMLOCK · 22
POLEMONIACEAE · 170, 171, 173, 174
POLYGONACEAE · 175, 177, 178
Polygonum · 175, 177
POPCORNFLOWER · 78
Populus · 204
Portulaca · 184

PORTULACACEAE · 184
Potentilla · 196
POVERTYWEED · 55, 105
PRICKLY PHLOX · 173
Prosartes · 142
Prunella · 135
Prunus · 197, 198
Pseudognaphalium · 62
Pseudostellaria · 98
Pseudotsuga · 161
Pterospora · 150
PUNTUREVINE · 222
Purshia · 198, 199
PURSLANE · 184
PUSSYTOES · 34

Q

Queen Anne's lace · 23

R

RABBITBRUSH · 44, 49
RAGWEED · 32
RAGWORT · 63
RANUNCULACEAE · 184, 185, 186, 188, 189, 190
Ranunculus · 189

RASPBERRY · 200
RED MAIDS · 179
REIN ORCHID · 157
RHAMNACEAE · 190
Rhus · 17
Ribes · 125
ROCKCRESS · 79
Rorippa · 91
Rosa · 199, 200
ROSACEAE · 191, 193,
    194, 195, 196, 197,
    198, 199, 200, 201,
    202
ROSE · 199
ROSE CAMPION · 97
RUBIACEAE · 203
Rubus · 200
Rudbeckia · 62
Rumex · 178
RUSCACEAE · 141

S

SAGE · 135
SAGEBRUSH · 37
SALICACEAE · 204,
    205
Salix · 205
SALSIFY · 66
Salsola · 106
SALTBUSH · 101
Salvia · 135
Sambucus · 94

SAND SPURRY · 100
SANDMAT · 115
SAND-VERBENA · 151
SANDWORT · 95, 98
SAPINDACEAE · 15
Saponaria · 99
Sarcobatus · 107
Saxifraga · 209
SAXIFRAGACEAE ·
    206, 207, 209
SAXIFRAGE · 209
SCOTCH THISTLE ·
    60
SCROPHULARIACEAE
    · 209, 210, 211, 212,
    213, 214, 215
Scutellaria · 136
SEDGE · 112
Sedum · 110
SELF-HEAL · 135
Senecio · 61, 63
SERVICEBERRY · 191
SHEPHERD'S PURSE ·
    83
Silene · 98, 99
Sisymbrium · 91
SKELETONWEED · 43
SKULLCAP · 136
SMARTWEED · 177
SNAKEWEED · 52
Snow Drops · 26
SNOWBERRY · 94
SOLANACEAE · 215
Solanum · 215, 216

Solidago · 63
Sonchus · 64
Sorbus · 201, 202
SPEARMINT · 134
SPEEDWELL · 215
Spergularia · 100
Sphaeralcea · 150
Spiderflower · 92
Spiraea · 202
SPIRAEA · 202
SPURGE · 115
SQUIRREL TAIL
    GRASS · 167
ST. JOHN'S WORT ·
    107
STARTHISTLE · 41
STARWORT · 98
Stellaria · 96, 98, 100
STICKSEED · 74
Stipa · 169
STONECROP · 110
STONESEED · 75
STORKS-BILL · 123
STRAWBERRY · 193
Streptopus · 143
SUMAC · 17
SUNFLOWER · 53
SWEETCICELY · 27
SWEETCLOVER · 120
SWEETROOT · 27
Symphoricarpos · 94
Synthyris · 213
Syringa · 126

T

TANSYMUSTARD · 86
Taraxacum · 31, 65
Tarragon · 38
TARWEED · 58
TEASEL · 112
Thalictrum · 190
THEMIDACEAE · 145
THIMBLEBERRY · 200
THISTLE · 45, 64, 106
Thysanocarpus · 92
TICKSEED · 47
TOADFLAX · 206
TOOTHWORT · 84
Toxicodendron · 18
Toxicoscordion · 144
Tragopogon · 66
Tribulus · 222
Trifolium · 120
Triteleia · 145
TRITELEIA · 145
TUMBLEMUSTARD ·
    91
TURKEY-MULLEIN ·
    115
TWISTED-STALK ·
    143
Typha · 216
TYPHACEAE · 216

U

ULMACEAE · 218
Urtica · 219
URTICACEAE · 219

V

VALERIANACEAE · 220
Veratrum · 147
VERBANACEAE · 220
Verbascum · 214
Verbena · 151, 220, 221
VERBENA · 220
Veronica · 215
VETCH · 121
Vicia · 121
Viola · 221
VIOLACEAE · 221
VIOLET · 221

W

WALLFLOWER · 87
WATER HEMLOCK · 20
WATERCRESS · 89
WATERLEAF · 126
Western Pearly-everlasting · 33

WHEATGRASS · 163, 167
WHITETOP · 85
WILD RYE · 167
WILLOWHERB · 155
WILLOWS · 205
WINTERCRESS · 80
WINTERFAT · 105
WOODLANDSTAR · 206
Woodsia · 113
WORMWOOD · 38
Wyethia · 39, 67
WYETHIA · 67

X

Xanthium · 68, 69

Y

YAMPAH · 27
YARROW · 31, 50
YELLOWCRESS · 91

Z

Zigadenus · 144
ZYGOPHYLLACEAE · 222

www.ingramcontent.com/pod-product-compliance
Lightning Source LLC
Chambersburg PA
CBHW071411170526
45165CB00001B/239